Understanding Energy Innovation

"Setting out to de-mystify energy innovation, this book provides a comprehensive, grounded and accessible overview of the insights that a social perspective on energy transitions brings. With a focus on smart grids, drawing on examples from Australia and around the world, it explores the dynamics of innovation in practice, the stories we tell about it, and how nostalgia for times gone past will shape energy futures. A practical, insightful guide for the transition pathways ahead."

—*Professor Harriet Bulkeley*, Department of Geography, Durham University, and Copernicus Institute of Sustainable Development, Utrecht University

"The electricity grid already was a uniquely complex machine so what does it take to make it 'smart'? Engineering can detail the material ingredients, but only the social sciences can explain the messy process of trying to make such innovations happen. In the short space of this unique book, Lovell provides expert guidance to the social science theories behind innovation, sheds new light on Australia's smart grid experiments and (wait for it) explains why nostalgia matters."

—*Dan van der Horst*, Professor of Energy, Environment and Society, School of GeoSciences, University of Edinburgh, UK

"Lovell presents an accessible and insightful framework for considering energy innovation. Through current case studies, she makes a powerful argument for more attention to be given to the social and human dimensions of innovation in the energy transition. This is a valuable contribution for those who commission and fund energy research, those who undertake research, and those who use the results."

—*Drew Clarke, Chair*, Australian Council of Learned Academies (ACOLA) Australian Energy Transition Research Plan

Heather Lovell

Understanding Energy Innovation

Learning from Smart Grid Experiments

Heather Lovell
School of Social Sciences & School
of Geography, Planning and Spatial Sciences
University of Tasmania
Hobart, TAS, Australia

ISBN 978-981-16-6252-2   ISBN 978-981-16-6253-9   (eBook)
https://doi.org/10.1007/978-981-16-6253-9

Cover pattern © Melisa Hasan

This Palgrave Macmillan imprint is published by the registered company Springer Nature Singapore Pte Ltd.
The registered company address is: 152 Beach Road, #21-01/04 Gateway East, Singapore 189721, Singapore

Acknowledgements

I would like to thank the Australian Research Council for funding the research that this book is based on, as part of their Future Fellowships programme (grant reference—FT140100646). I would also like to thank all those who generously gave their time to participate in the research, from households to senior executives and policy specialists. I am sincerely grateful to colleagues at the University of Tasmania who provided their research support and expertise at different times over the course of my Future Fellowship grant, in particular Dr Phillipa Watson and Dr Cynthia Nixon. Also, to Ruth Steel for her immense help with the final stages of this book.

A big thank you to Palgrave Macmillan and particularly to Josh Pitt for his help and encouragement throughout.

Finally, thanks to my family—Francisco, Nico, Isabel and Emily.

CONTENTS

LIST OF FIGURES

CHAPTER 1

Introduction

No one talks about smart grids any more. Smart grids were popular a
few years ago. If I were you I would research something else.
—Interview, Energy Consultant, April 2015

This response to my planned research was from my first interviewee. I had just arrived in Australia from the other side of the world, family in tow, to start a four-year social research project on smart grids. Understandably, I was a little dismayed. But I knew for a fact that there were lots of smart grid projects and ideas around, and it was interesting that some saw smart grids as unfashionable. How do we make these sorts of judgements about policy and technology fashions, and what influences them? What are the popular stories that circulate about smart grids, and what do those stories leave out? These early encounters in smart grid research only confirmed what I already suspected, that smart grids are as much about society as they are about technology: smart grids are inherently social.

In *Understanding Energy Innovation*, I draw on new empirical findings from a four-year project on the social science of smart grids, including over fifty research interviews and several focus groups and workshops. My primary focus is Australia, but I also incorporate case studies from Europe and North America. The book is about the process of digital or smart innovation within the electricity sector, with a focus on the social and political. Simply defined, smart grids are the incorporation of new digital

© The Author(s) 2022
H. Lovell, *Understanding Energy Innovation*,
https://doi.org/10.1007/978-981-16-6253-9_1

and information and communication technologies into utility infrastructures.

Understanding Energy Innovation sits alongside a host of technical books on smart grids and utility innovation and is deliberately quite different from these. The ambition here is to embrace, celebrate and learn from the messiness of social responses to smart grids, rather than ignore, curtail or curb society in order to facilitate smart grid implementation. In other words, the social aspects of energy sector innovation are celebrated and placed centre stage.

WHO THE BOOK IS FOR AND HOW TO READ IT

In this book, I explore and explain energy innovation using smart grids as a case study. Energy innovation is something many of us are trying to get a better handle on as we grapple with climate change, high energy prices, unreliable supply, and the emergence of new technologies. I present a number of ways to think about and plan for energy sector reform and innovation, drawing on core ideas from social and innovation theory. I write about these theoretical ideas in an accessible, jargon-free way, recognising that a diversity of people have an interest in energy innovation generally, and in smart grids more specifically, and would like to find out more about different ways of understanding energy innovation from a social science perspective.

The book is intended to meet a growing demand for learning about social research among energy sector professionals in engineering, computer science, economics, and utility planning. I have observed growing interest from these professions in social research, having had the opportunity to work closely with industry and government on several collaborative smart grid projects over the past decade. *Understanding Energy Innovation* is also intended for academics and university (undergraduate and postgraduate) students. It is likely to be of particular interest to interdisciplinary postgraduates studying planning, energy studies, energy and society, environmental science, and environmental engineering.

The book can be read from cover to cover, but it is also designed to be picked up and put down. Each of the four main chapters—on networks, nodes, narratives and nostalgia—works as a stand-alone text, and the key ideas and learnings from each chapter are summarised in the conclusion.

Aims and Themes

My main goal with this book is to de-mystify social responses to innovation and show how unexpected things will happen with any innovation project. In the case of smart grids, many unforeseen events happened when smart grids were implemented, and the same interventions turned out differently in different places. The same is true for energy innovation more generally. The book uses the example of smart grid experiments to explore and explain energy innovation processes and to summarise learnings from smart grid experiments in several countries at the forefront of smart grid innovation.

Several detailed case studies are presented from countries where smart grid developments have rapidly advanced in the last decade, with a focus on Australia, Europe, and North America. The optimistic promise of smart grids is contrasted against what happened in practice with smart grid implementation. Technologies go wrong, budgets escalate, installers are not properly trained or are rushed, and the values and preferences of householders and other electricity customers are forgotten. Throughout the book, smart grids are used to illustrate wider energy sector innovation ideas, issues, and processes. From the politics of framing a smart grid project a failure, to the ways that new knowledge about smart grid technologies and ideas circulate globally and are tested in different places, the book takes a social science perspective to explore and ultimately celebrate the complex sociotechnical processes of energy innovation.

The four core themes of the book, which aim to capture key aspects of energy sector innovation from a social perspective, are **networks**, **nodes**, **narratives**, and **nostalgia**. These themes bring together relevant scholarship on technology innovation from human geography, science and technology studies, political science, and sociology. Each theme is explored under a separate chapter designed to work as a stand-alone text, introducing the key ideas and examining them through three short case studies.

- **Networks** explores the many different types of network that social scientists study to better understand processes of change, from policy networks to sociotechnical networks. I draw on the well-understood and familiar example of an electricity network to shine a light on how a myriad of networks exist in society, albeit with different types of link between the components of each network. Some networks are primarily technical (like electricity networks), and some primarily social, but all have elements of both the technical and the social. Networks also all have in common a number of characteristics such as interconnectedness, flows, network-wide effects, and fragility. A focus on networks helps us better understand the social nature of energy innovation. I explore case studies on international smart grid policy networks, a local community network on Bruny Island, Australia, and a fragile network—a digital metering programme in the State of Victoria, Australia.

- **Nodes** are closely related to networks, as nodes are fixed, stable passage points on networks. On an electricity network, a node is an electricity meter, an inverter, or a substation. But electricity network nodes can also be social things that operate in ways that provide predictability and order, such as organisations. A focus on nodes helps us to understand the ways in which stability is provided within complex networks, and how stability can quickly erode as the role of a node changes or there is a malfunction. The case studies presented here are about different types of nodes: the digital electricity meter, with a focus on household transitions in the UK and Australia; an electricity regulator, in this case, the Australian Energy Market Operator; and islands as nodes, with a focus on King Island, Australia.

- **Narratives** centre on the stories that circulate in society that help us simplify and make sense of innovations such as smart grids. Narratives are helpful to study not only because of the things, people and organisations they speak to but also because of the things not spoken about—the silences. There are many narratives about smart grids. In the case studies, I explore a global industry narrative about households and their willingness to participate in smart grids; multiple narratives about a smart grid project in the State of Victoria, Australia; and a case of competing narratives of energy futures, examining off-grid and the hydrogen economy narratives.

- **Nostalgia** is about a longing for the past: the way we remember how things used to be done and a wish for things to stay the same. Although nostalgia is pretty much the opposite of innovation, and therefore perhaps does not immediately seem relevant to this book, it is actually a central part of understanding social responses to energy sector innovation because every new technology and way of doing something is in effect competing with nostalgia. I examine how nostalgia can hamper efforts at smart grid innovation, particularly in how it blinds us to change already underway, but also how positive memories can act to encourage innovation. I analyse the effects of nostalgia by drawing on three diverse case studies: memories of pioneering international smart grid experiments and their present-day effect; a case study of the phenomenon of scarce data on off-grid households in Australia, which is shown to be linked to nostalgia; and nostalgia for *big infrastructure* revealed in tensions in planning for the future of the electricity grid in Australia.

THEORIES OF ENERGY INNOVATION

There are lots of different concepts and theories to help us understand energy innovation. In this book, I focus on social perspectives: approaches that take society seriously and place people, organisations, values, and cultures centre ground. Social perspectives are distinct from the technical engineering and economic analyses of the energy sector that tend to dominate policy and industry discussion. These are about uptake rates, cost curves, value stacks, demand curves, and so on. While these topics are undoubtedly important, they only go so far in explaining energy innovation.

The main social theories about innovation in the energy sector—and utility infrastructures more generally—can be roughly categorised according to the scale of analysis and the issue or thing they are focusing on. Here I group the theories into two camps: people-technology interaction, and people-focused. I briefly summarise a large amount of academic scholarship on these theories below. If you wish to dive into the theory more, I recommend that you follow up on some of the references at the end of the chapter.

People-Technology Interaction

People-technology interaction innovation theories relevant to the energy sector cover two main topics: innovation in large-scale sociotechnical systems (electricity networks, transport infrastructures, gas networks), and small-scale human-technology interactions.

Large-scale sociotechnical system theories are generally about change over decades, such as how we moved centuries ago from a transport system based on horse-drawn carriages to one with motor cars (Bridge et al., 2018). They draw on research findings from historical examples. These sociotechnical theories provide a structure for thinking about how and why innovation occurs and the patterns of change. They help us understand by categorising different aspects of the innovation process. The main type of categorisation used across these theories is to do with the scale of change: from initial small-scale niche testing of new ideas and technologies, to diffusion up to regime level change, and finally, the broadest scale—enduring landscape level changes. The core idea is that innovations progress from the niche scale to the other levels over time. There are several slightly different variants of this scale-based or diffusion-over-time sociotechnical theory, including the multi-level perspective, strategic niche management, and large technical systems theory (Hughes, 1983; Kemp et al., 1998; Markard et al., 2012; Smith et al., 2010).

Small-scale human-technology interaction theories are primarily about the household and everyday energy technology interactions in the home. This area of research puts the household centre stage, exploring household behaviours, habits, and values, alongside the material and technical energy infrastructure of the home (such as type of housing material, technology used for heating and cooling or fuel type). It is a well-established area of innovation research, which dates back to the energy crisis of the 1970s (Guy & Shove, 2000; Hinchliffe, 1996). The study of the *prosumer*—households who both *pro*duce and con*sume* energy—is a more recent focus (Parag & Sovacool, 2016).

In household human-technology research, the person and the technology are studied with equal attention. This method is called symmetry because it is about equal (symmetrical) attention to people and technologies (Callon, 1986; Murdoch, 1997). This approach has mostly been used in an area of research called actor-network theory. The detail of actor-network theory is quite involved, and there is not space to explain it here (see Latour, 2005 for a good summary), but it is about the relationships between humans and technologies, particularly the processes by which these relationships change and stabilise (see Chap. 2 for an example of its application).

A second area of research about household habits and patterns relating to energy technologies is social practice theory (Shove et al., 2012). A core idea of this area is that households are not deliberately consuming energy. Instead, they consume energy *services*, heating, lighting and so forth. These services could be provided in a number of different ways to households, using different technologies and different fuels, but with the energy service level remaining the same from the householder's perspective. Therefore, to understand household demand for energy services, pretty much every aspect of day-to-day life in the household is relevant: what time people get up, what their washing and bathing habits are, whether there is someone at home during the day, and so on. There is a lot of diversity in households in terms of the energy infrastructure and how homes are made, as well as household preferences and ways of doing things. This type of detailed social research is fundamentally important to understanding energy innovation because there is a tendency in the energy sector to oversimplify and overgeneralise the response of householders to a new energy intervention (such as the installation of a household battery or solar panels) and assume that all households will react in pretty much the same way (Ellabban & Abu-Rub, 2016). Several studies have shown this is not the case, with considerable diversity between households (Bulkeley et al., 2016; ECA, 2020; Ransan-Cooper et al., 2020).

People-focused

People-focused theories are not directly about technologies or the energy sector. The ideas and approaches I use include discourse analysis, the study of narratives, and the study of memories and nostalgia. Although they stem from different social science disciplines, these concepts, at their core, are all about narratives. They are about stories concerning the present, past and future (including how these stories are remembered) that influence how we think about and act in relation to energy sector innovation. These approaches have been used in political science to understand change in policies and ways of governing (see e.g., Dryzek & Schlosberg, 1998) and in organisation studies to understand change (Czarniawska, 1997). These theories are distinct from the people-technology theories described above because they do not consider the role of technologies. I delve into these ideas in more detail in the chapters on narratives and nostalgia.

DEFINING SMART GRIDS

Here I provide some background on smart grids, as smart grids are the case study used throughout the book to explore energy innovation. If you already know a lot about smart grids, you may want to skip straight to the next chapter.

Smart grid concepts and practices have been applied to a range of utility modernisation projects in transport, water, and gas. But most smart grid activity to date has been with electricity, and electricity is the focus of this book. So from here on, the term *smart grid* is used to denote electricity smart grids unless otherwise explained. Smart grids are initiatives that involve the digitalisation of the electricity sector: the application of new computer science techniques and technologies to the electricity grid with the aim of improving its function. The US National Institute of Standards and Technology defines smart grids as:

> the addition and integration of many varieties of digital computing and communication technologies and services with the power-delivery infrastructure. (NIST, 2014, p. 33)

And energy researchers expand on this as:

> the modernisation of the electricity-delivery system to allow for greater automation in grid operation at virtually every node, including facilitating data communications and operations between all agents in the system, which include generators, system operators, and final demanders (consumers). (Guo et al., 2015, p. 7)

These definitions are typical in terms of putting the technical aspects of smart grids front and centre, with social objectives and the role of society given less emphasis. The graphic of a smart grid below (Fig. 1.1) is pretty typical of smart grid illustrations, which tend not to have any people in them.

Some definitions of smart grids are more people-focused, such as this one from a European Commission Task Force:

> A Smart Grid is an electricity network that can cost efficiently integrate the behaviour and actions of all users connected to it—generators, consumers and those that do both—in order to ensure economically efficient, sustainable power system with low losses and high levels of quality and security of supply and safety. (EU Commission Task Force for Smart Grids, 2010, p. 6)

As this definition makes clear, smart grids have emerged in response to several policy problems: rising electricity prices, intermittent supply, and environmental sustainability. In this book, I examine some of these policy drivers (see, e.g., Case Study 4.1), but the focus is more on what happened with smart grid *implementation*. I note too that the majority of empirical material in the book is about Australia, and the policy drivers in Australia for smart grids have mostly not been about environmental

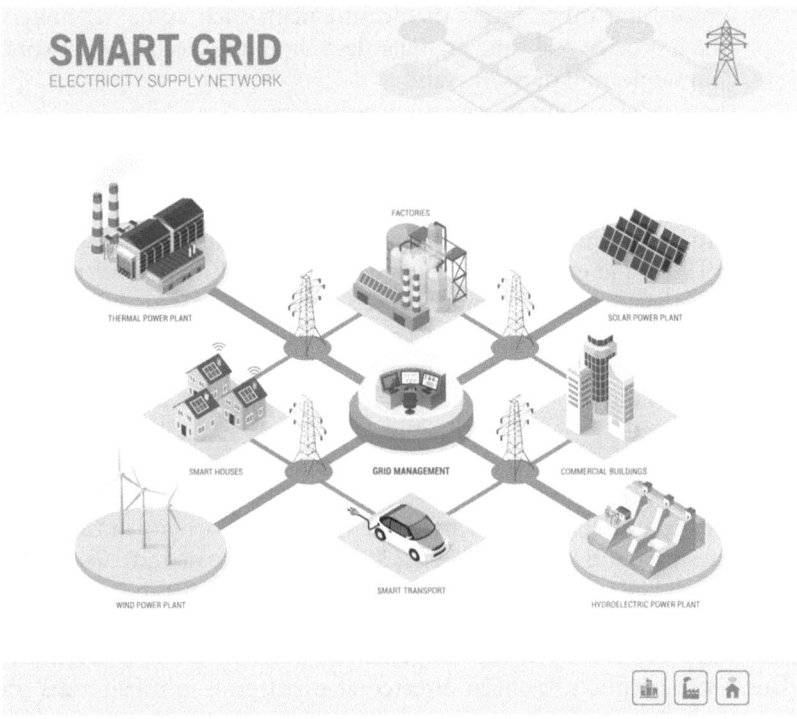

Fig. 1.1 Illustration of an electricity smart grid. (Source: iStock)

problems (readers looking for an environmental analysis of smart grids might refer to Evans et al. (2019) or Gabrys (2014)).

Although, in general terms, *smart* usually means the integration of new digital capabilities into existing utility infrastructures (smart grids), confusingly, there are also smart cities, intelligent networks, smart homes and so on. Smart grid is also a bit of a catch-all term that brings together a range of different types of digital technology innovation in electricity networks. This adds further confusion because a project might be described as a smart grid but might incorporate different innovations to a smart grid elsewhere. Smart grids generally include one or more of the following:

- new energy generation technologies such as solar photovoltaics and other renewable forms of generation;

- sensors and other forms of measurement, such as digital meters, which provide real-time granular data about the electricity network;
- communication networks; and
- forms of electricity storage, such as batteries.

A Brief History of Smart Grid Innovation

The term smart grid first emerged and started gaining popularity during the late 1990s and early 2000s, when new digital technologies were being developed that allowed for a much more responsive grid system. Electricity meters—positioned as they are at the interface of the customer and the grid—were a particular focus in the early stages of smart grid innovation. It was the country of Italy that was at the forefront of innovation with their Telegestore project. Telegestore began in 1999 and was completed by 2006; it involved the installation of 32 million smart meters across Italy (ISGAN, 2019).

The period of 2008 to 2013 saw the most global activity in the use of the term smart grid (Google Trends, 2019). Since the mid-2010s, the use of smart grid as a term has waned in popularity somewhat, but this partly reflects the growth of other similar terms such as smart city. Smart grid and smart city are often used together, or interchangeably, because to date, most attention has been directed at smart grids in urban areas (de Jong et al., 2015). Urban areas are viewed as hotspots of innovation where the capital required for smart grids (such as finance and human resources) is present and where lots of utility infrastructures come together.

Smart grids are an expression and outcome of the rise of the internet and digital technologies in society. The core types of professionals involved in smart grids are computer scientists and information and communication technology experts. So, smart grids bring together the traditional energy profession of electrical and power system engineering with computer science. Smart grids aim to improve and modernise energy infrastructure. The electricity networks that have provided countries in Europe, North America, and Australia with centralised, reliable electricity have remained pretty much unchanged for almost a century. Here smart grid innovation is being driven by the availability of new information and communications (digital) technology but also by the ageing of energy infrastructures and the need for substantial repairs, upgrades and investment.

Several new energy policy problems have also emerged since our electricity grids were first built, that smart grids are seen as a solution to. The main problems have been described as the energy trilemma, the three interlinked problems of achieving energy security, mitigating climate change, and ensuring energy affordability (Bradshaw, 2013). For instance, climate change is increasing residential air conditioning requirements in many countries as temperatures rise. This extra demand from households requires better management of substantial afternoon and evening peaks in demand from the grid. In many countries, households are increasingly installing solar photovoltaic panels and other forms of electricity generation, which often require extra investment in the local grid to ensure the excess electricity fed into the grid does not adversely affect its operation. Investment in our electricity infrastructures are pushing up electricity prices in many places, which particularly affects low-income households, exacerbating energy poverty (ACOSS, 2018).

Even from this short summary, it is evident that there are a wealth of policy problems facing the energy sector and the electricity sector within it. Smart grids are one among a range of energy innovations being touted and tested in response to these problems.

In this book, I focus mainly on Australia, where I conducted most of my fieldwork on smart grids. Australia is a good place to study smart grids because it is one of the countries that has been at the forefront of electricity grid innovation internationally. This is because of early government investment in smart grids, digital meters, and household solar photovoltaics (rooftop solar). The growth of rooftop solar was incentivised by state government initiatives in Australia during the 2000s and encouraged by high electricity prices and a sunny climate. Just over a fifth of Australian households currently have rooftop solar (DISER, 2020). Because of Australia's high proportion of household prosumers, the electricity grid has had to be adjusted to cope with the new two-way flows of electricity, not only *to* households but also *from* households. It is no surprise that household battery companies consider Australia as a key market.

Smart grids are an example of a policy initiative that promised a lot initially and came with much optimism and excitement. In the Australian context, the potential of smart grids has been described in very optimistic terms, for example, in a government strategy document about smart grid standards:

> The optimal deployment of smart grids holds significant potential for the management of many of the challenges confronting the electricity supply chain in Australia. (Lazar & McKenzie, 2012, p. 1)

and by a government minister:

> Smart grids represent the cutting edge of energy efficient technologies, applied in energy production, distribution and householder use, a frontier the Australian Government is committed to exploring quickly and strategically as we move to a low-carbon future. (Australian Government Minister for the Environment, Heritage and the Arts, cited in DEWHA, 2009, p. 4)

As smart grids have become more widespread and have increased in popularity, some social scientists have criticised the lack of clarity about what smart actually means in practice. That is, how *smart* has been translated from a vision into something implemented on the ground. As the sociologist Hollands explains in relation to smart cities, "the disjuncture between image and reality here may be the real difference between a city actually being intelligent, and it simply lauding a smart label." (2008, p. 305). This disjuncture between narrative and reality is something worth paying attention to with smart grids, as in practice, much of the original promise of smart grids has not necessarily been realised when they have been implemented (Lovell, 2019). This book explores how we can better understand this gap between the promise of smart grids and the reality.

Social scientists have also questioned the close (perhaps too close) alignment between smart grids and corporate interests, noting how:

> Smart technologies may provide innovative ways to reduce carbon, decentralise energy generation, and provide security from external threats, **but once they are released into the 'real world' they can become co-opted by corporate interests and subsumed under existing power relations.** (McLean et al., 2016, p. 3253, emphasis added)

Such analysis adopts a critical stance to the smart ideology focus on economic efficiency, market function and business opportunities, with a lack of attention to issues of social equity and environmental sustainability. Identifying critical social and environmental justice issues within the optimistic (and typically business-orientated) narratives about smart grids has been a central plank of social science smart grids research to date (Evans et al., 2019).

WHY USE SMART GRIDS AS A WAY TO UNDERSTAND ENERGY INNOVATION?

Smart grids are a good example of how what might seem like a purely technical initiative is, in fact, deeply social. As a manager of an energy advocacy organisation illustrates in his description of smart metering programmes (a central element of smart grids):

> At the start we thought that this was a technical reform but what we realise now is that it was actually a social reform. And in treating it as a technical reform we got the social side of it wrong. That's a common refrain that keeps coming up. (Interview, May 2015)

Smart grids also reflect something we often see across the energy sector, which is the mismatch between planned (often aspirational) objectives and the realities of implementation. In other words, how society has responded to smart grids is quite often a long way away from how those involved in smart grids thought it would at the planning stage. Overall, smart grid projects have taken longer to implement, have cost more, and have had fewer financial benefits than expected. There are lots of sensible explanations for why smart grid initiatives have not always worked as planned. Society is very diverse, there are multiple different interests in smart grids (including strong corporate interests and profit motives), and there are lots of new technically uncertain things and unknown risks. So, perhaps the better question is why anyone would ever think that a new smart grid project would run smoothly or be implemented and work in the same way everywhere.

To understand better the mismatch between smart grid objectives and actual implementation, we need to look to society. This is what this book is about: exploring the messiness of social responses to smart grids so that those of us involved in smart grids and other energy innovation projects can proceed with our eyes wide open.

REFERENCES

ACOSS. (2018). *Energy stressed in Australia*. Australian Council of Social Service (ACOSS). Retrieved February 10, 2021, from https://www.acoss.org.au/wp-content/uploads/2018/10/Energy-Stressed-in-Australia.pdf

Bradshaw, M. (2013). *Global energy dilemmas*. Polity.

Bridge, G., Barr, S., Bouzarovski, S., Bradshaw, M., Brown, E., Bulkeley, H., & Walker, G. (2018). *Energy and society: A critical perspective*. Routledge.

Bulkeley, H., Powells, G., & Bell, S. (2016). Smart grids and the constitution of solar electricity conduct. *Environment and Planning A, 48*(1), 7–23.

Callon, M. (1986). Some elements in a sociology of translation: Domestication of the scallops and fishermen of St. Brieuc Bay. *Sociological Review Monograph, 32*(2), 196–233.

Czarniawska, B. (1997). *Narrating the organization: Dramas of institutional identity*. University of Chicago Press.

de Jong, M., Joss, S., Schraven, D., Zhan, C., & Weijnen, M. (2015). Sustainable-smart-resilient-low carbon-eco-knowledge cities; making sense of a multitude of concepts promoting sustainable urbanization. *Journal of Cleaner Production, 109*, 25–38.

DEWHA. (2009). *Smart grid, smart city: A new direction for a new energy era*. Department of the Environment, Water, Heritage and the Arts, Commonwealth of Australia.

DISER. (2020). *Solar PV and batteries*. Australian Government Department of Industry, Science, Energy and Resources (DISER). Retrieved March 3, 2021, from https://www.energy.gov.au/households/solar-pv-and-batteries

Dryzek, J. S., & Schlosberg, D. (1998). *Debating the Earth: The environmental politics reader*. Oxford University Press.

ECA. (2020). *Power shift: Final report*. Energy Consumers Australia (ECA). Retrieved February 20, 2020, from https://energyconsumersaustralia.com.au/wp-content/uploads/Power-Shift-Final-Report-February-2020.pdf

Ellabban, O., & Abu-Rub, H. (2016). Smart grid customers' acceptance and engagement: An overview. *Renewable and Sustainable Energy Reviews, 65*, 1285–1298. https://doi.org/10.1016/j.rser.2016.06.021

EU Commission Task Force for Smart Grids. (2010). *Expert Group 1: Functionalities of smart grids and smart meters*. Retrieved January 13, 2021, from http://www.ieadsm.org/wp/files/Tasks/Task%2017%20-%20Integra tion%20of%20Demand%20Side%20Management,%20Energy%20Efficiency, %20Distributed%20Generation%20and%20Renewable%20Energy%20 Sources/Background%20material/Eg1%20document%20v_24sep2010%20 conf.pdf

Evans, J., Karvonen, A., Luque-Ayala, A., Martin, C., McCormick, K., Raven, R., & Palgan, Y. V. (2019). Smart and sustainable cities? Pipedreams, practicalities and possibilities. *Local Environment, 24*(7), 557–564.

Gabrys, J. (2014). Programming environments: Environmentality and citizen sensing in the smart city. *Environment and Planning D: Society and Space, 32*(1), 30–48.

Google Trends. (2019). *Google trends 'explore' – Search term 'smart grid'*. Retrieved September 1, 2019, from https://trends.google.com/trends/explore?date= 2004-01-01%202019-09-03&q=smart%20grid

Guo, C., Bond, C. A., & Narayanan, A. (2015). *The adoption of new smart-grid technologies: Incentives, outcomes, and opportunities*. RAND Corporation.

Guy, S., & Shove, E. (2000). *A sociology of energy, buildings and the environment: Constructing knowledge, designing practice*. Routledge.

Hinchliffe, S. (1996). Helping the earth begins at home: The social construction of socio-environmental responsibilities. *Global Environmental Change, 6*(1), 53–62.

Hollands, R. G. (2008). Will the real smart city please stand up? *City, 12*(3), 303–320.

Hughes, T. P. (1983). *Networks of power: Electrification in Western society 1880–1930*. The John Hopkins University Press.

ISGAN. (2019). *AMI CASE Case05 – ITALY*. International Smart Grid Action Network (ISGAN). Retrieved September 10, 2019, from http://www.iea-isgan.org/ami-case-case05-italy/

Kemp, R., Schot, J. W., & Hoogma, R. (1998). Regime shifts to sustainability through processes of niche formation: The approach of Strategic Niche Management. *Technology Analysis and Strategic Management, 10*(2), 175–195.

Latour, B. (2005). *Reassembling the social: An introduction to actor-network-theory*. Oxford University Press.

Lazar, J., & McKenzie, M., 2012. *Australian Standards for Smart Grids – Standards Roadmap*. A report for the Federal Department of Resources, Energy and Tourism, Canberra, Australia.

Lovell, H. (2019). The promise of smart grids. *Local Environment, 24*(7), 580–594.

Markard, J., Raven, R., & Truffer, B. (2012). Sustainability transitions: An emerging field of research and its prospects. *Research Policy, 41*(6), 955–967.

McLean, A., Bulkeley, H., & Crang, M. (2016). Negotiating the urban smart grid: Socio-technical experimentation in the city of Austin. *Urban Studies, 53*(15), 3246–3263.

Murdoch, J. (1997). In human/nonhuman/human: Actor-network theory and the prospects for a nondualistic and symmetrical perspective on nature and society. *Environment and Planning D: Society and Space, 15*(6), 731–756.

NIST. (2014). *NIST framework and roadmap for smart grid interoperability standards, release 3.0*. National Institute of Standards and Technology. Retrieved June 5, 2021, from https://www.nist.gov/system/files/documents/smart-grid/Draft-NIST-SG-Framework-3.pdf

Parag, Y., & Sovacool, B. K. (2016). Electricity market design for the prosumer era. *Nature Energy, 1*(4), 1–6.

Ransan-Cooper, H., Lovell, H., Watson, P., Harwood, A., & Hann, V. (2020). Frustration, confusion and excitement: Mixed emotional responses to new household solar-battery systems in Australia. *Energy Research & Social Science, 70*, 101656.

Shove, E., Pantzar, M., & Watson, M. (2012). *The dynamics of social practice: Everyday life and how it changes.* Sage.

Smith, A., Voß, J.-P., & Grin, J. (2010). Innovation studies and sustainability transitions: The allure of the multi-level perspective and its challenges. *Research Policy, 39*(4), 435–448.

CHAPTER 2

Networks

WHAT IS A NETWORK, AND WHAT TYPES OF NETWORK DO WE FIND IN SMART GRIDS?

A network, put simply, is a group formed from parts that are connected together (Cambridge Dictionary, 2020) or a collection of points joined together in pairs by lines (Newman, 2018, p. 1). Networks describe relationships between things—how things and people are linked together. There are many different types of network, from networks of humans (social networks) to ecosystems of plants and animals, and networks within and between organisations. Many different types of network have been observed across the Earth's ecosystems, infrastructures, and societies, from food system supply chains, to 5G communication networks, to social networks. In relation to energy sector innovation, most networks comprise a mix of social and technical elements: sociotechnical networks are about people-technology interactions (see Chap. 1 for a summary of academic theories in this area). We see evidence of people-technology interactions in the three smart grid examples discussed in this chapter.

© The Author(s) 2022
H. Lovell, *Understanding Energy Innovation*,
https://doi.org/10.1007/978-981-16-6253-9_2

There is overlap between the terms system and network, and they are often used interchangeably. For instance, Thomas Hughes, a historian of technology, defines a system in a very similar way to the network definition above as "constituted of related parts or components" (Hughes, 1983, p. 5), but for Hughes, the network is the physical or material structure that links the components. In science and technology studies, the term system is used more often than network to describe large complex infrastructures such as electricity. An alternate view is to see networks as a simplification of systems, as the physicist Newman explains:

> A network is a simplified representation that reduces a system to an abstract structure or topology, capturing only the basics of connection patterns and little else… a lot of information is usually lost in the process of reducing a full system to a network representation. (Newman, 2018, p. 7)

It is therefore important to remember that in order to study networks, things and people that are judged to be less critical to the operation of the network (and/or to the interest of the person studying the network) are excluded from the conceptualised network. This curation means that the boundaries of any conceptualised network are imposed or idealised; in practice, they tend to be much more blurred and fluid.

CHARACTERISTICS OF NETWORKS AND THEIR RELEVANCE TO SMART GRIDS

The key characteristics of networks are their components and the relationships between them. In other words, networks are about multiple components and the relations between these components, which, in turn, define how the network is arranged and how it functions. There are many ways that networks can be analysed, such as calculating how centralised the network is, looking at subgroups within a network, or identifying powerful nodes. Many of these techniques are quantitative and use mathematical theories. Here I concentrate on the qualitative study of networks. The main reason networks are studied is because the pattern of interactions within a network can affect network behaviour or outcome. In other words, understanding network interactions helps us to understand network function.

Networks are relevant to smart grids because, first and foremost, smart grids are about the digitalisation of (in most parts of the world) an already existing significant infrastructure network: the electricity grid. At its heart, the large-scale electricity grid system is about connectivity: providing electricity to multiple consumers from large-scale electricity generators. Traditionally it has been a supply network, although, in recent decades, there has been a shift towards a more complex grid to also manage generation from consumers (prosumers) because of decentralised electricity generation. Smart grids involve modernising existing infrastructure through the integration and overlay of new digital technologies and capabilities. This allows the traditional (physical) grid infrastructure to be operated more effectively because there is real-time data on factors such as generation, consumption, voltage, and condition of the electricity lines.

In the above description, the electricity network is described mainly as a physical network of technologies: power lines, sensors, substations. There are many examples of definitions of the electricity network that follow this kind of description, for instance, the following from industry association Energy Networks Australia:

> Electricity Network means transmission and/or distribution systems consisting of electrical apparatus which are used to convey or control the conveyance of electricity between generators' points of connection and customers' points of connection. (ENA, 2008, p. 3)

This definition is very technical, focusing on the material infrastructure and objects that comprise the electricity network. In reality, there are many fundamentally social aspects of an electricity network—from the rules and standards that define the technical specification of the network, to the decision-making of end consumers. Where the smart grid gets interesting, and considerably more complicated, is when we consider the human actors within smart grid networks. As mentioned in the introduction to this book, often people are quite absent from smart grid definitions, as in this definition from the US National Institute of Standards and Technology:

> [smart grids involve] the addition and integration of many varieties of digital computing and communication technologies and services with the power-delivery infrastructure. (NIST, 2014, p. 33)

Humans are, of course, relevant to smart grids in a number of ways, whether as individuals or as part of households, organisations or community groups. If we start to see smart grids as mixed social and technical (sociotechnical) networks, this is a bit more complicated, but it better reflects the actual on-the-ground workings of smart grids. It is when smart grids are conceptualised as just a technical network that we can run into problems, as a manager of an energy NGO described to me in relation to the implementation of a smart metering programme:

> A lot of the leadership of the project had been handed over [by the government] to consultants. And it really felt to me like there was a lack of understanding in government, that **it was not just an industry issue but that it was a public policy, social policy and political issue** that needed leadership that reflected all of that. And I think they got there in the end but it was painful to get there. (Interview, November 2016)

Different Ways of Thinking About Networks

In the biological sciences, the main focus of network analysis is plants, animals, and ecosystems. In computer science, attention in recent years has concentrated heavily on machine learning networks. In the social sciences, policy networks are an area of interest—groups of actors from inside and outside of government that together influence what is on the policy agenda and how well policies are implemented. Social network analysis is another key area of social science research that uses quantitative data on who knows who to create intricate maps showing social relations. Social scientists studying technology are interested in networks made up of people and technologies: sociotechnical networks. They aim to take a neutral view about which type of actor in the network is doing work, technology, or person. In other words, there is an openness to non-human things—devices, infrastructure, technology, computers—doing equivalent work to humans. Sociotechnical networks are a key type of network analysed in this book because of their strong relevance to energy innovation. Table 2.1, below, provides a summary of the types of networks that social scientists are interested in, and how different disciplines conceptualise and explain the diverse functions of these networks.

Table 2.1 Types of networks studied by social scientists

Type of network	Field / discipline	Common terms and descriptors	Core characteristics	Definitions
Policy network	Political science	Networks, coalitions, policy network analysis, policy communities	• Comprised of people, organisations, values, beliefs and resources • Structures (institutions) that define, generate and implement policies • Operate as a form of collaborative governance, including both formal and informal, public and private interactions • Develop over time and tend to be longstanding, but are also constantly evolving • Change typically happens in response to changes outside of the policy network (exogenous changes)	"Policy networks are sets of formal institutional and informal linkages between governmental and other actors structured around shared if endlessly negotiated beliefs and interests in public policymaking and implementation. These actors are interdependent and policy emerges from the interactions between them." (Rhodes, 2006, p. 424) "[Policy] Networks… result from repeated behaviour and, consequently, they relieve decision makers of taking difficult decisions; they help routinize behaviour. They simplify the policy process by limiting actions, problems and solutions." (Marsh & Smith, 2000, p. 6) "Operating in a more or less institutionalized setting, [policy network] actors are engaged in horizontal, relatively nonhierarchical discussions and negotiations to define policy alternatives, or formulate policies, or implement them." (Coleman, 2001, p. 11608)

(*continued*)

Table 2.1 (continued)

Type of network	Field / discipline	Common terms and descriptors	Core characteristics	Definitions
Sociotechnical network	Science and technology studies, innovation studies	Actor-networks, sociotechnical systems, large technical systems, technological networks	• Comprised of social and technical things, which become coherent actors (i.e., capable of doing a task or performing a service) if they are closely interconnected • Technologies and other material things provide stability to the network, especially if they are durable (e.g., power stations) • Characterised as unstable in actor-network theory but stable within sociotechnical system theories	"Sectors like energy supply, water supply, or transportation can be conceptualized as socio-technical systems. Such systems consist of (networks of) actors (individuals, firms, and other organizations, collective actors) and institutions (societal and technical norms, regulations, standards of good practice), as well as material artifacts and knowledge…. The systems concept highlights the fact that a broad variety of elements are tightly interrelated and dependent on each other…. This has crucial implications for the dynamics the systems exhibit, and especially for system transformation." (Markard et al., 2012, p. 956) "networks are comprised of diverse materials, woven together in order to ensure the durablity of the consolidated relations…such relations count for little unless they are held together by durable and resilient materials." (Murdoch, 1998, p. 360) "networks are not **in** the actors, but are produced **by** them." (Callon, 1991, p. 155)

| Economic / business networks | Economic geography, business studies | Global Production Networks, network models | • Comprised of corporations and their products
• Strong international focus
• Networks of production and consumption
• Business networks are specifically about the relationship of one business with others (their network), differing from other network theories, which are about "the relations between all elements of the set" (Kilkenny & Fuller-Love, 2014, p. 303); that is, they include technologies or other types of organisation.
• Economic networks operate at multiple scales and levels of analysis (Knoke, 2014)
• Emphasis on quantitative analysis (e.g., quantitative network models are used to assess network structures and function (see Jackson, 2010)) | Global Production Networks are "the complex firm networks and territorial institutions involved in globalized economic activity, and how these are structured both organizationally and geographically." (Coe & Yeung, 2019, p. 777)
"Economic networks operate at multiple levels of analysis, including individuals (consumers, employees), groups (households, work teams), organisations (firms, interest groups), and populations (industries, markets), as well as across these levels." (Knoke, 2014, p. 3) |
| Social networks | Sociology, economic sociology, organisation studies | Social network analysis, social networks | • Comprised of humans
• Emphasis on detailed *social network analysis* using computer software and quantitative techniques
• The technique has been used to study a range of different social networks, from religious groups to friendship patterns | "A social network is a set of socially relevant nodes connected by one or more relations. Nodes, or network members, are the units that are connected by the relations whose patterns we study. These units are most commonly persons or organisations." (Marin & Wellman, 2011, p. 11) |

One way to think about characterising different types of network is by what links the components of the network. For instance, in international smart grid networks, it is expertise, work and organisations that link members (see Case Study 2.1). In community smart grid networks, it is typically location and a sense of place that links the members (see Case Study 2.2 of Bruny Island). In other spheres of life, it might be values or beliefs that are the link binding the members of a network together. For example, the political scientist Sabatier's ideas about *advocacy coalitions* focus on networks of people working across a range of organisations (public and private) who are united to push for policy change based on their values (Sabatier & Jenkins Smith, 1993).

Network resilience depends on the strength of these linkages. A theory within science and technology studies called actor-network theory (see Chap. 1; also Table 2.1) is relevant to the discussion because of its focus on early-stage innovations. Actor-network theory is one of the people-technology interaction theories that study both the social and the technical elements of networks. A key observation is how fragile sociotechnical networks are: they are very prone to breaking down. One of the founders of actor-network theory, Michael Callon, demonstrates in his case of scallop conservation in France how when the scallop larvae failed to thrive in new specially designed collector units, and they were harvested too early by the fishermen, the network failed (Callon, 1986). From Callon's classic case to examples in agriculture (Higgins & Kitto, 2004), housing (Lovell & Smith, 2010) and medicine (Singleton & Michael, 1993), actor-network theory scholars have examined issues of network fragility, breakdown and failure.

According to Callon and others, a critical stage in the development of new networks is that of *translation*—a process by which previously disparate things and people are brought together into a coherent network (an actor-network) that can act in a unified way. Callon and other science and technology studies scholars have noted the amount of work involved in translation, as well as the ongoing effort required to maintain network stability (Callon, 1986; Murdoch, 1997). These ideas about the tendency of sociotechnical networks to disintegrate have also been applied to utility infrastructures (Graham & Marvin, 2001; Sovacool et al., 2018). This is highly relevant in thinking about smart grids and other types of energy innovation, where lots of hidden behind the scenes work goes into keeping things like electricity networks running smoothly. As Graham and Marvin in their book *Splintering Urbanism* remind us, infrastructure networks are "precarious achievements" (2001, p. 182).

CASE STUDY 2.1 INTERNATIONAL SMART GRID
POLICY NETWORKS

Italy decides to implement new digital electricity meters in all households, thirty two million in total, a programme that commences in 1999 and requires new meters to be specially built, as none yet exist. The State of Victoria in Australia decides in early 2006 to upgrade its planned new electricity metering programme to an advanced metering programme so that meters can communicate with each other (DPI, 2007). In a suburb of Austin, Texas, a comprehensive smart grid pilot is implemented in 2008 (The Pecan Street Project, 2010). How are these diverse decisions and programmes connected? The answer: through international smart grid policy networks. No smart grid pilot or initiative is done in isolation. People in government, corporations, and other types of organisation working in the field of smart grids do not make decisions alone, but rather with reference to what has gone before, and in other places.

Several international governance organisations facilitate this very type of information exchange. There is the International Smart Grids Action Network through which governments share smart grid ideas, policies and programmes (ISGAN, 2015). There is also an equivalent corporate international group—the Global Smart Grid Forum (recently renamed as the Global Smart Energy Federation) (GSEF, 2020). Mission Innovation is another international organisation that has a dedicated smart grid programme based on sharing learning between countries and specific sites of smart grid innovation (Mission Innovation, 2021). There are also international standards organisations such as the International Organisation of Legal Metrology and the International Electrotechnical Commission that work closely to set new standards to underpin smart grids and to facilitate seamless smart grid operation from country to country.

Through these international policy networks, smart grid activities worldwide are connected, reviewed and learnt from. To more fully understand the social aspects of smart grids and their effects, it is essential to recognise that these international networks exist and are active in sharing both the good and the bad of smart grid practices. These global networks bring both advantages and disadvantages (more about this later).

A branch of social science called policy mobilities seeks to better understand this type of international policy network (Peck & Theodore, 2010). Policy mobility scholars are interested in how policy ideas and programmes have joined the globalisation bandwagon. The idea is that with globalisation, policies are becoming increasingly mobile—travelling around the world and being implemented in different places, based on what one country or city has been doing over in Asia or Europe. The international policy networks mentioned above are pivotal in facilitating this global diffusion of smart grid policies from country to country, city to city. We can see many examples of this. For instance, the graph below (Fig. 2.1) shows how Australian smart grid projects have been referenced internationally from a sample of over one hundred international policy and industry reports.

We can also drill down to look at specific smart grid projects and how other countries have cited them. Data on two large Australian smart grid projects—Smart Grid Smart City and the Victorian Advanced Metering Infrastructure (AMI) Program—is shown in the graph below (Fig. 2.2). Interestingly, there was a peak in referencing these projects in 2011 to 2012 at the beginning of the implementation of Smart Grid Smart City and partway through the implementation of the Victorian AMI Program. This peak in interest was much higher than when the findings and data were released at the end of the projects (2014 and 2013, respectively).

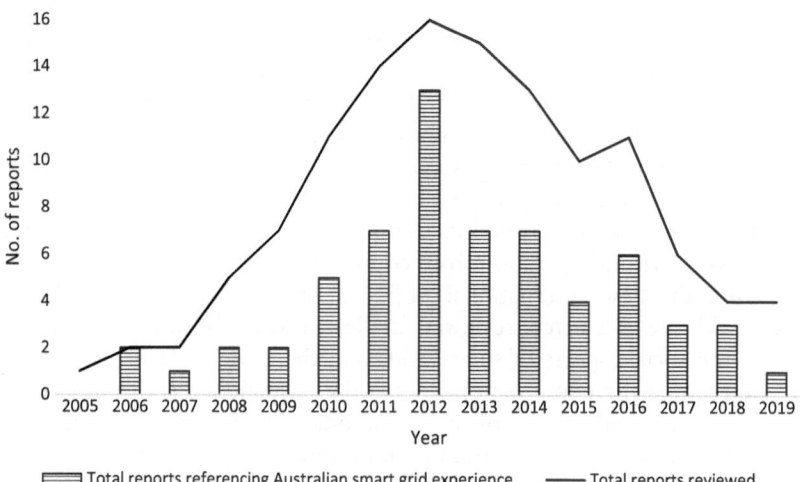

Fig. 2.1 Number of international policy documents referencing the Australian smart grid experience over time. (From analysis provided by Dr Cynthia Nixon, University of Tasmania)

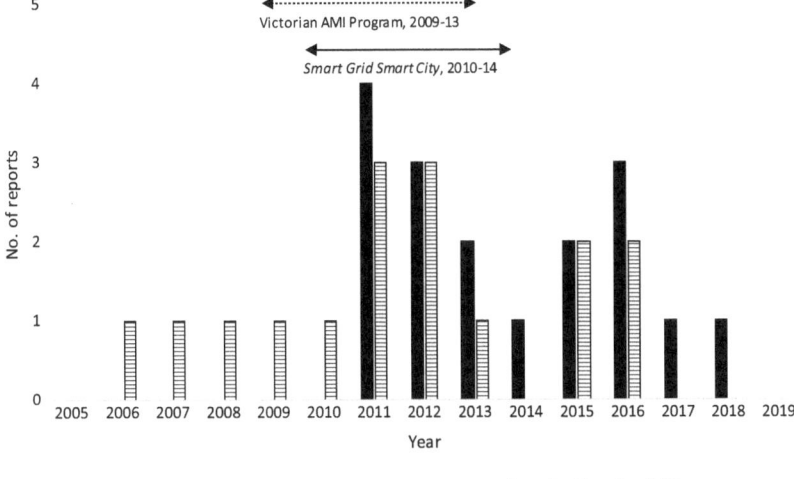

Fig. 2.2 References to the Smart Grid Smart City Project and the Victorian Advanced Metering Infrastructure (AMI) Program in international documents over time. (From analysis provided by Dr Cynthia Nixon, University of Tasmania)

In turn, Australia has drawn heavily on other countries to learn about smart grids. For example, the main policy document produced at the outset of the Australian Smart Grid Smart City project explains how:

> Smart Grid, Smart City can also take lessons from other domestic and international smart grid related initiatives, such as Solar Cities in Australia; Smart Grid City in Boulder, Colorado, US; other US smart grid demonstration projects; and the PRIME project in Europe. (DEWHA, 2009, p. 12)

However, there is also some caution about the applicability of lessons from overseas, as the International Electrotechnical Commission smart grid standardization roadmap notes:

> The power distribution system in the USA, Canada and many other countries of the world (Brazil, Mexico, Australia, South Africa, Korea etc.) is significantly different to the distribution system in Europe. (IEC, 2017, p. 54)

This caution is demonstrated too in a comment made to me by an innovation manager of an Australian utility, at the time heavily involved in a large smart grid demonstration project:

So I guess one of the things the AER [Australian Energy Regulator] always try to do is compare us as networks because 'a pole's a pole's a pole' but we do have quite different geographical constraints across the [Australian] States. In the latest determination they've been comparing us to Canada as well which obviously is extremely different. (Interview, April 2015)

From this brief review, it is clear that international policy networks can encourage or hinder innovation, depending on how programmes and ideas from elsewhere are interpreted.

CASE STUDY 2.2 A LOCAL COMMUNITY NETWORK: BRUNY ISLAND, AUSTRALIA

Bruny Island sits just off the south-east coast of another Australian island, Tasmania. It has a modest population of around 800 people, which surges significantly in the summer due to an influx of tourists and second home (shack) owners. Bruny Island is connected to the primary electricity grid in Tasmania through two undersea cables. While Bruny Island is mostly a story about a community (social) network, the undersea cables are a key component of this network. The cables are old and cannot supply quite enough electricity to Bruny Island to meet peak demand. For several years, TasNetworks, the local utility, has had to run diesel generators at times of peak demand to compensate.

In 2015 a group of universities, along with TasNetworks, the local utility, and an Australian energy start-up company, Reposit Power, were awarded funding for an alternative solution to meet peak demand, rather than using the diesel generators. This was a smart grid solution, with household battery storage installed in around thirty households on Bruny Island, along with rooftop solar. Electricity was automatically drawn from these batteries at times when the electricity network required it. I was part of this project team, leading the social research component of the project, along with four other social researchers.[1] The rest of the team were engineers, economists, and information and communication technology experts.

At the start of this three-year smart grid project, the physical, material features of Bruny Island and its electricity system were the main focus of

---

[1] Including Dr Phillipa Watson, Dr Andrew Harwood, and PhD student Veryan Hann (all of University of Tasmania) as well as Dr Hedda Ransan-Cooper, Australia National University.

discussion within the project team: the undersea cable, the technical inter-face between the household batteries and the grid, the operation of the diesel generator and so on. But by the end of the project, it was just as likely that social issues were a point of discussion in our regular project meetings: whether it was the latest findings from our social research, or the high volume of rumour mill driven inquiries TasNetworks was receiv-ing from households. During the project, it became apparent that there was a tight-knit community on Bruny Island and a community whose preferences around energy storage were quite different from those previ-ously observed in urban areas. This presented a challenge within the proj-ect team because Reposit Power had developed their product largely within a metropolitan area in Canberra, Australia. Most of their experience was with urban and urban-fringe households.

Bruny Island was distinctive in that many of the households in the smart grid trial had concerns about being left without power during energy outages. Outages are frequent on Bruny Island because it is an edge-of-grid location, where the electricity network is more expensive and difficult to maintain. Plus, Bruny Island has many highly forested areas with power lines prone to being damaged by tree falls, as you can see in the photograph below of one of the island's roads (Fig. 2.3). During the trial, there was one outage of three days that affected many trial house-holds. Because of this context of frequent, sustained electricity grid out-ages, trial households were understandably interested in the role their battery could play in providing back-up power to their household. Back-up power was a key issue raised by Bruny Islanders right from the start of the project. Perhaps none of this sounds especially surprising, but it was sur-prising to Reposit Power. Reposit Power does not offer back-up power to households as part of its product. Instead, its product is focused on pro-viding a service to utilities based on the certainty of being able to draw power from household batteries when utilities require it.

An illustration of the value that the Bruny Island trial households placed on back-up power was that most paid several hundred dollars extra to have their batteries provide this type of emergency power. This cost is not insig-nificant given that the median income of households on Bruny Island is A$34,000 a year, well below the Australian national average (A$45,000) (ABS, 2015).

The case of Bruny Island gives us insights into the role that community can play in influencing the implementation of smart grid technologies. This was a close-knit community and one with particular vulnerabilities to

Fig. 2.3 Long rural road on Bruny Island, Australia, showing trees near power lines. (Source: Dr Phillipa Watson, University of Tasmania)

electricity network outages. The community, therefore, had a particular expectation of what household level battery storage meant for them. In other words, the context on Bruny Island strongly affected the smart grid trial—both positively and negatively (Watson et al., 2019). Similar findings about the importance of context have been identified by Laura Watts in her energy social research on the Orkney Islands, Scotland. As she writes poetically of Orkney electricity and its relationship to the communities on Orkney:

> The Orkney electron… has both electrical and political power… it is constituted by islander people and their engagement, who make it brighten and flow. (Watts, 2018, p. 72)

On Bruny Island, the local community raised issues that were not expected by the research team. These issues were helpful for project learning and may indicate what will occur during the implementation of other island smart grids, or those in similar remote and rural locations. But what we also see in the case of the Bruny Island smart grid is the interplay between the social and technical. It is this complex sociotechnical network of issues and things, from household anxieties about outages to technical product specifications, that is important to recognise and attend to in order to better understand processes of energy innovation.

CASE STUDY 2.3 A FRAGILE NETWORK: THE STATE OF VICTORIA'S DIGITAL METERING PROGRAM, AUSTRALIA

The State of Victoria's mandatory digital (or advanced) metering programme is a good example of a smart grid network breaking down. Victoria was Australia's first state to privatise its electricity sector in the late 1990s. Victoria was keen to proceed with digital metering so that its newly privatised market could function better: digital meters facilitate greater choice of consumer tariffs and easier household switching of electricity company provider. So, in 2004 state government approval was given in Victoria to proceed with an interval metering programme. Interval meters were an early-stage digital meter that collect consumption data in a digital form but do not transmit or communicate the data remotely. As the technology choice in meters rapidly improved in the mid-2000s, the Victorian metering programme was upgraded to advanced meters in 2006. Advanced meters have communications embedded and so can transmit data remotely, without having to be manually read. The Victorian Advanced Metering Infrastructure (AMI) programme ran from 2009 to 2013 and resulted in 2.3 million digital meters being installed in 93% of homes and small businesses in Victoria.

However, not everything went smoothly with the implementation of digital meters in Victoria: the network unravelled. This was partly because of changes to components of the AMI programme network but also to do with the changing context in which the network was operating. A key component that changed was the AMI network's star performer: the digital meter. The late 2000s and early 2010s were a period of rapid innovation for digital meters internationally. The State of Victoria was testing these new technologies at scale, which inevitably led to some teething problems. As Adrian Clark, Head of Smart Metering Australia at Landis Gyr, an international metering company, explained, "the Victorian problems emanated from decisions taken almost 10 years ago, and since that time the [metering] technology has leapfrogged" (cited in MacDonald-Smith, 2015).

Technical teething problems with the meters were resolved, but the issues delayed implementation and pushed up costs; costs which were passed down to households and small businesses. Rising costs became a source of tension with the AMI programme. There were also technical issues with the communication systems that the new digital meters relied on to provide remote reading of electricity consumption and other services. Again, this was a rapidly developing area of technology over the four to five year period of the AMI programme, so lots of things were being learned and refinements made during the programme implementation, rather than at a pilot or prototype stage. The instabilities in the core technical components of the AMI network created further instabilities across the network. This is in keeping with actor-network theory, which highlights the ability of technologies to either disrupt or stabilise networks.

There were also changes in the wider context in which the AMI programme network was situated. Most notably, in 2010, at a crucial stage in the life of the AMI programme (as implementation was just getting underway), there was a change of government in the State of Victoria, from Labor to a Liberal-National government. The new government had campaigned in the election on the AMI programme, raising the possibility that it would stop it. This was because problems were already starting to emerge with the programme. There was growing public discussion about some of these, particularly the high metering implementation charge to households, many of whom had not yet had a meter installed. However, the new government did decide, somewhat reluctantly, to proceed, albeit with notable changes to the programme, including introducing optional flexible pricing, establishing a Ministerial Advisory Council, and subsidising in-home energy displays (see Victorian State Government, 2015). The new State Energy Minister explained the decision as follows:

> analysis shows that if you were looking at it from a blank sheet of paper you probably wouldn't go down this [AMI program] path. There are actually more detriments to consumers, or costs to consumers as the result of the project as a whole, compared to the benefits. But we're not starting with a blank sheet of paper. We're starting with the mess we've inherited from the Labor government. (Victorian Energy Minister Michael O'Brien, 2011)

As we know from how international smart grid policy networks operate (see Case Study 2.1), new information was also continuously flowing in from other countries about different ways of implementing meters. In particular, the case of New Zealand's metering programme came to be important because this was a voluntary implementation programme; that is, households did not have to get a new digital meter. In New Zealand the programme was opt-in or market-led, and so was quite different to

Victoria's mandatory implementation approach. New Zealand was frequently cited as a counterbalance to the negative case of Victoria, as an Australian state government manager explained at the time:

New Zealand is largely seen as a positive example and Victoria as a negative one. (Interview, April 2015)

The State of Victoria's AMI programme network was fragile because so many things kept changing—the components of the network were unstable—including its core technology: the digital meter.

LEARNING FROM SMART GRID NETWORKS

Smart grids demonstrate how energy sector networks are a mix of the social and the technical, that is, sociotechnical networks. Like all energy systems, the successful function of smart grids is achieved not only in efficient technical operation but through the whole network of things and people working together in harmony. The smart grid case studies presented in this chapter highlight the diversity of networks of different technologies, materials, people and organisations that drive energy innovation. In the table below, I summarise the key learnings from these smart grid network case studies and suggest how they might guide future practice.

Key learning	Recommendation for energy practitioners
Smart grids are sometimes conceived of as technical networks, whereas in reality, they are sociotechnical (part social, part technical).	When energy innovations such as smart grids are conceptualised as just a technical network, we can run into problems: from the outset, energy sector innovations should ideally be thought about as much as a social program as a technical one.
Decisions about energy innovations are not made in isolation—there are international policy networks that continuously circulate new ideas and information, and these information flows can have both positive and negative effects.	Participating in international policy networks is beneficial, but knowledge shared within these networks can often be quite edited (i.e., the most positive narrative of what happened). Connecting directly with the people about projects to also find out what went wrong is likely to provide more detail than is found in the version presented in international policy network discussions.
There are many different types of network relevant to smart grids and energy innovation: policy, social, sociotechnical, and business.	The existence of multiple networks means that implementing new technologies is most likely not going to be straightforward. Steps to mitigate technology risk such as piloting the technology or building in review periods are important, as is managing expectations.

References

ABS. (2015). *Australian Bureau of Statistics (ABS) – Bruny Island – Kettering (SA2) (603021069) population data.* Retrieved March 14, 2018, from https://itt.abs.gov.au/itt/r.jsp?RegionSummary&region=603021069&dataset=ABS_REGIONAL_ASGS&geoconcept=REGION&measure=MEASURE&dataset ASGS=ABS_REGIONAL_ASGS&datasetLGA=ABS_NRP9_LGA&regionLGA=REGION&regionASGS=REGION

Callon, M. (1986). Some elements in a sociology of translation: Domestication of the scallops and fishermen of St. Brieuc Bay. In J. Law (Ed.), *Power, action, belief* (pp. 196–233). Routledge and Kegan Paul.

Callon, M. (1991). Techno-economic networks and irreversibility. In J. Law (Ed.), *A Sociology of Monsters: Essays on power, technology and domination* (pp. 132–161). Routledge, London & New York.

Cambridge Dictionary. (2020). *Network – definition.* Retrieved February 12, 2021, from https://dictionary.cambridge.org/dictionary/english/network

Coe, N. M., & Yeung, H. W.-c. (2019). Global production networks: Mapping recent conceptual developments. *Journal of Economic Geography, 19,* 775–801.

Coleman, W. D. (2001). Policy Networks. In N. J. Smelser & P. B. Baltes (Eds.), *International Encyclopedia of the Social & Behavioral Sciences* (pp. 11608–11613). Pergamon.

DEWHA. (2009). *Smart Grid, Smart City: A new direction for a new energy era.* Department of the Environment, Water, Heritage and the Arts, Commonwealth of Australia.

DPI. (2007). *Victorian government rule change proposal – Advanced metering infrastructure rollout.* Department of Primary Industries (DPI).

ENA. (2008). *List of definitions.* Energy Networks Australia (ENA). Retrieved December 18, 2020, from https://www.energynetworks.com.au/assets/uploads/energy_networks_australia_list_of_definitions_0.pdf

Graham, S., & Marvin, S. (2001). *Splintering urbanism: Networked infrastructures, technological mobilities and the urban condition.* Routledge.

GSEF. (2020). *About GSEF (Global Smart Energy Federation).* Retrieved June 1, 2015, from http://globalsmartenergy.org/page/186/global-smart-grid-federation

Higgins, V., & Kitto, S. (2004). Mapping the dynamics of new forms of technological governance in agriculture: Methodological considerations. *Environment and Planning A, 36,* 1397–1410.

Hughes, T. P. (1983). *Networks of power: Electrification in Western society 1880–1930.* The John Hopkins University Press.

IEC. (2017). *International Electrotechnical Commission: Smart grid standardisation roadmap.* Retrieved December 20, 2020, from https://webstore.iec.ch/publication/27785

ISGAN. (2015). *About us – what is ISGAN?* International Energy Agency. Retrieved May 25, 2015, from http://www.iea-isgan.org/?c=1

Jackson, M. O. (2010). *Social and economic networks.* Princeton University Press.

Kilkenny, M., & Fuller-Love, N. (2014). Network analysis and business networks. *International Journal of Entrepreneurship and Small Business, 15*(21), 303–316.

Knoke, D. (2014). *Economic networks.* John Wiley & Sons.

Lovell, H., & Smith, S. J. (2010). Agencement in housing markets: The case of the UK construction industry. *Geoforum, 41*(3), 457–468.

MacDonald-Smith, A. (2015, June 21). Landis+Gyr eyes growth opportunity in smart meter roll out. *Sydney Morning Herald.* http://www.smh.com.au/business/energy/landisgyr-eyes-growth-opportunity-in-smart-meter-roll-out-20150617-ghpx1i.html#ixzz3r3rdA3DH

Marin, A., & Wellman, B. (2011). Social network analysis: An introduction. *The SAGE Handbook of Social Network Analysis, 11*, 25.

Markard, J., Raven, R., & Truffer, B. (2012). Sustainability transitions: An emerging field of research and its prospects. *Research Policy, 41*, 955–967.

Marsh, D., & Smith, M. (2000). Understanding policy networks: Towards a dialectical approach. *Political Studies, 48*, 4–21.

Mission Innovation. (2021). *Mission innovation: IC1 smart grids.* Retrieved March 5, 2021, from http://mission-innovation.net/our-work/innovation-challenges/smart-grids/

Murdoch, J. (1997). In human/nonhuman/human: Actor-network theory and the prospects for a nondualistic and symmetrical perspective on nature and society. *Environment and Planning D: Society and Space, 15*(6), 731–756.

Murdoch, J. (1998). The Spaces of Actor-Network Theory. *Geoforum, 29*, 357–374.

Newman, M. (2018). *Networks.* Oxford University Press.

NIST. (2014). *NIST framework and roadmap for smart grid interoperability standards, release 3.0.* National Institute of Standards and Technology. Retrieved June 5, 2021, from https://www.nist.gov/system/files/documents/smartgrid/Draft-NIST-SG-Framework-3.pdf

Peck, J., & Theodore, N. (2010). Mobilizing policy: Models, methods, and mutations. *Geoforum, 41*(2), 169–174.

Rhodes, R. A. (2006). Policy network analysis. *The Oxford Handbook of Public Policy*, 432–447.

Sabatier, P. A., & Jenkins Smith, H. C. (Eds.). (1993). *Policy change and learning: An advocacy coalition approach.* Westview Press.

Singleton, V., & Michael, M. (1993). Actor-networks and ambivalence: General practitioners in the UK cervical screening programme. *Social Studies of Science, 23*(2), 227–264.

Sovacool, B. K., Lovell, K., & Ting, M. B. (2018). Reconfiguration, contestation, and decline: Conceptualizing mature large technical systems. *Science, Technology, & Human Values, 43*(6), 1066–1097.

The Pecan Street Project. (2010). *The Pecan Street Project – Working group recommendations.* Retrieved February 12, 2016, from https://www.smartgrid.gov/files/documents/Pecan_Street_Project_Report_Recommendation_201012.pdf

Victorian Energy Minister Michael O'Brien. (2011). *Smart meters here to stay despite cost blow-out.* Retrieved November 9, 2015, from http://www.abc.net.au/news/2011-12-14/smart-meter-roll-out-continues-despite-cost-blow-out/3730522

Victorian State Government. (2015). *Smart meter government review and decision.* Retrieved October 30, 2015, from http://www.smartmeters.vic.gov.au/about-smart-meters/government-review

Watson, P., Lovell, H., RansanCooper, H., Hann, V., & Harwood, A. (2019). *Consort Bruny Island battery trial project final report – Social Science.* Report Produced for Australian Renewable Energy Agency (ARENA). https://arena.gov.au/assets/2019/06/consort-social-science.pdf

Watts, L. (2018). *Energy at the end of the world: An Orkney Islands Saga.* MIT Press.

CHAPTER 3

Nodes

What Is a Node, and What Types of Node Do We Find in Smart Grids?

Nodes are parts of networks where elements of the network intersect. They are at junction points of flows—a point of intersection or an *obligatory passage point* (Callon, 1986). The most basic definition of a node is a point where one thing joins another (Collins Dictionary, 2021) or a point in a network or diagram at which lines or pathways intersect or branch (Google Definitions, 2021). Within smart grids, nodes can be organisations, people, or technologies. Nodes are anchor points and typically act as brokers at critical intersections within smart grids. Nodes are therefore best seen as social or technical, or a mix of both, sociotechnical. Either way, nodes are important in smart grids in terms of providing stability, order, and co-ordination.

Nodes can be many different things. In the social sciences, nodes are usually individuals or organisations. In the physical sciences, nodes are typically material things, such as sensors or inverters, and in the biological sciences, they are junctions within plant structures or circulatory systems. The characteristics of nodes are closely related to the network they are part of: some networks have many nodes, and some only have a few. The position of the node on a network affects the agency the node has, and the network outcomes (Borgatti et al., 2009). Nodes are sites worthy of attention because many things happen when flows intersect at nodes. Nodes can operate well, in which case they are usually not noticed very much, or they can become sites of blockage and malfunction. If nodes malfunction, they tend to attract a lot of attention; think of a traffic light that breaks down, causing a traffic jam.

Different Ways of Thinking About Nodes

There are many different terms used to describe nodes across different research areas including broker (e.g., knowledge brokers or policy brokers), boundary object, and junction (e.g., innovation junction). Node is the default term used in biology and the physical sciences. Scientists in areas such as botany and computer science use the term node to describe the intersection of parts of plants (plant nodes) and junctions in machine learning models. A branch of geography—economic geography—uses the term node to describe large cities, such as London and New York, which are viewed as important global sites of innovation and trade (Sassen, 2002). Node is also the primary term used within the social sciences in social network analysis. Social network analysis is a methodology and approach to analyse patterns within social networks; a node is an individual or organisation within the network. The bigger the node, the more social linkages that person has (Borgatti et al., 2009).

Elsewhere in the social sciences, nodes are more commonly referred to using other terms, including brokers, entrepreneurs, boundary objects and junctions. Brokers are individuals who have particular skills in connecting people and exchanging ideas and information. Knowledge broker is a popular concept used across several social science disciplines, including business and management studies and sociology. A knowledge broker is defined by the sociologist of innovation Morgan Meyer as "people or organizations that move knowledge around and create connections between researchers and their various audiences." (Meyer, 2010, p. 118). Policy broker (sometimes referred to as policy entrepreneur) is a term used in the political sciences to describe individuals adept at connecting different policy networks. The political scientist John Kingdon (2003, p. 122) first coined the term *policy entrepreneur* to describe people who are "[willing] to invest their resources—time, energy, reputation, and sometimes money—in the hope of a future return", with the *return* in this instance being policy change. A number of authors have subsequently drawn on Kingdon's work to explore how individuals generate new ideas and catalyse change within government, acting as important nodes in the policy system (see e.g., Bartlett & Dibben, 2002; Etzkowitz & Gulbrandsen, 1999).

The term boundary object was first introduced by science and technology studies scholars Star and Griesemer (1989) to explain how a natural history museum functions as a node. The museum enables different social groups involved in the museum's work—scientists, field ecologists, university administrators, farmers and animal trappers—to work together effectively to collect, classify and analyse specimens. In this context, boundary objects were things like classification systems, specimens, field notes, or maps of particular territories (1989, p. 408), which all had some potential to be differently understood by the different social groups that worked for the museum. Although this example is obviously far removed from the energy sector, it can usefully be applied in terms of considering the distinct types of expertise operating relatively discretely within the energy sector—from economists to planners to power engineers—each with slightly different understandings of, and perspectives about, what energy sector innovation is.

The boundary object concept is used in this chapter to describe the changing role of electricity meters in the home (see Case Study 3.1). A boundary object is defined as a node positioned between different social groups or types of organisation, such as householders, government, and utilities. A boundary object has a lot of flexibility in how it is understood, allowing useful work to be done even in situations where there is conflict or misunderstandings between different social groups. A focus on boundary objects is highly relevant to smart grids and energy innovation more generally, where there are many entities (typically technologies or organisations) that play the role of mediating between different social groups.

Innovation junctions are bounded spaces in which multiple technologies are being used. The core idea is that grouping multiple technologies in one particular place—the junction or node—leads to innovation. De Wit et al. use the office as an example of an innovation junction. In their historical analysis of changing office technologies in the Netherlands from 1880 to 1980, they define the innovation junction as:

> a space in which different sets of heterogeneous technologies are mobilized in support of social and economic activities and in which, as a result of their co-location, interactions and exchanges among these technologies occur. These interactions and exchanges lead to location-specific innovation patterns. (de Wit et al., 2002, p. 50)

In the following case studies, I consider three different types of smart grid nodes: an electricity meter, an organisation, and an island.

Case Study 3.1 The Digital Electricity Meter as a Node: Household Transitions in the UK

Over the last decade or more there have been programs world-wide to replace traditional mechanical spinning disc meters with new digital meters (see also Case Study 2.3). Governments and energy sector organisations have framed the digital meter as an important new node in future energy systems. They see digital meters as critical to the transition to a secure, low carbon energy system at an affordable cost and in allowing for greater engagement and interaction with householders (GSGF, 2012; ISGAN, 2014). As the UK government agency with responsibility for the uptake of smart meters describes:

> The smart meter roll-out programme constitutes the largest transformation of a core area of nationwide infrastructure undertaken in a generation. It is a programme that aims to reach every household across the whole of Great Britain. (Smart Energy GB, 2013, p. 8)

With the transition to digital, the energy meter has expanded from the simple function of measuring energy demand to a host of features, mostly aimed at the householder, such as providing detailed feedback on energy consumption from digital devices and facilitating new types of energy tariff. Through multiple programs to replace meters in people's homes, utilities and governments have made meters a focal point of action and discussion for energy innovation. The meter is an important node that sits at a vital point in the electricity network, between utilities and consumers (households, businesses), and helps to shape relationships between them.

The idea of a boundary object (briefly introduced above) helps us to understand the ways in which a meter acts as a node. As we saw with the example of how workers in a natural history museum relate differently to the objects in the collections, so the digital meter is understood differently by the social groups of government, utilities, and consumers (Lovell et al., 2017). In comparison, traditional mechanical energy meters are a long-standing, relatively stabilised technology. The flexibility of how the digital meter can be interpreted has had both positive and negative effects. These effects are demonstrated well in the example below, the UK digital metering program, and a more detailed case study of an intensive community energy project involving meters.

In the UK, digital metering has been managed through a voluntary program, with overall targets set, but with the idea of encouraging households to obtain digital meters rather than making it compulsory. The UK government program commenced in 2016, and the aim is to have fifty three million new meters in place by the now extended deadline of 2024. The deadline was initially set at 2020 but was delayed in late 2019, largely in response to the COVID-19 pandemic (see Ambrose, 2019). Progress to date has been slower than expected but is still continuing reasonably well, with just under twenty four million smart meters in place across the UK as of the end of 2020 (UK BIES, 2020).

Part of the explanation for the relatively high digital metering uptake by UK households is how the UK government took the time and effort to better understand the social world of householders in the initial design of the program (Lovell et al., 2017). For example, a compulsory feature of the UK program is a digital home interface (DECC, 2012), which allows households to easily access and interpret digital metering data so that they can receive good quality real-time feedback on their consumption. Making the digital home interface compulsory was a decision that came out of prior research with households, seeking to understand how households value digital meters from their own perspective. The case of the UK smart metering program shows how the incorporation of household perspectives into the program design has facilitated the implementation of new digital meters—mediating between government, utilities and households—and reducing the amount of conflict between these different social worlds.

I was involved in a research project looking at changes to metering in Wyndford (Hawkey & Webb, 2014; Hawkey et al., 2016), a community in the west of the city of Glasgow, Scotland, that has high unemployment and high levels of socio-economic deprivation (Scottish Government Statistics, 2011). A new district heating system was implemented in Wynford in 2012, providing space heating and hot water for approximately 2000 homes. The upgrade was organised by the local housing association (Cube Housing Association) and the utility (SSE), with part funding from a Community Energy Saving Programme award from British Gas. As part of the district heating, new digital meters were installed in every home.

The meters were very different to what Wyndford householders had previously. At the same time as the metering change, the tariffs in Wyndford changed from pay as you go (i.e., a card-based top-up system, so households only had power if they were in credit) to a new tariff with a standing charge and usage charge that was averaged out over the year. The intention of the new tariff was to reduce bill shock during the winter months when heating requirements are, of course, much higher. But many households found the changes in tariffs very difficult to manage on a tight household budget. This was exacerbated by the new digital meter, which showed the daily charge plus levelized (annual average) consumption over a year and not actual consumption data. The organisations involved in the new district heating thought that households would not be interested in any more detailed data than this, and had located a second meter showing this data out of reach in an inaccessible place. A group of households complained about the new metering set up and associated tariffs. Changes were subsequently made, including providing access to the second consumption data meter and dropping the daily standing charge for some households.

This brief look at digital meters highlights how they act as a node at a critical boundary between the household and other organisations, including utilities and government. Meters control, standardise, and frame the identities of, and relationships between, the social worlds of government, utilities, and householders. Energy meters are currently in the midst of a contested and uneasy transition period, with old framings and ways of doing shifting into new ones. This reframing of the meter, and the relationships it shapes and mediates, are issues that the boundary object concept helps us to understand. By focusing on the meter itself as a node, changing social practices and relations are usefully brought to the fore.

CASE STUDY 3.2 THE AUSTRALIAN ENERGY MARKET OPERATOR AS AN ENERGY INNOVATION NODE

The Australian Energy Market Operator (AEMO) plays a crucial role in overseeing the market function of electricity and gas markets and hence acts as an important node in energy innovation. Over the past few years, it has developed a twenty-year plan for Australia's energy sector, called the Integrated System Plan (ISP). The ISP has become key to how energy innovation and transition are framed and discussed in Australia. AEMO describes the electricity sector (NEM: National Electricity Market) as a fast-changing complex system:

The NEM is an intricate system of systems, which includes regulatory, market, policy and commercial components. At its centre is the power system, which is an inherently complex machine of continental scale. This system is now experiencing the biggest and fastest transformational change in the world since its inception over 100 years ago. (AEMO, 2020, p. 10)

Such a complex system of systems requires organisational nodes to enable good governance. Through the ISP, AEMO has become an important node or broker in facilitating dialogue and decision-making about Australia's energy futures on a long-term, twenty-year horizon. AEMO has acted as a node of innovation in Australia, co-ordinating multiple stakeholders' inputs and carefully analysing system changes and risks. AEMO is similar to organisations internationally, such as the National Grid Electricity System Operator in the UK and the California Independent System Operator in the US (see National Grid ESO, 2020).

The Australian Commonwealth, state and territory governments established AEMO in 2009 to manage the National Electricity Market (NEM). The NEM operates in most states in Australia but not the Northern Territories. In 2015, AEMO's remit was extended to include the State of Western Australia (which has its own separate electricity network, not connected to the NEM). AEMO also now looks after gas markets. AEMO is an independent body jointly owned by governments (60%) and market participants (40%). AEMO has three main areas of responsibility: maintaining secure electricity and gas systems, managing gas and electricity markets, and leading the design of Australia's future energy system. It is this third function that is of most interest in relation to energy innovation. The shift to long term planning in Australia is in response to the increasing pace of change in its electricity system. A manager at AEMO explained back in 2015 the problems they were facing in this regard:

one could almost say that at some point in the future you may need to put sell-by dates on the advice [we produce at AEMO]. The National smart meter specification was good to a point, but then it became outdated and then you have to move on and it's … the market is moving so what we've just spoken about may be largely irrelevant in another five years. (Interview, Manager, AEMO, April 2015)

In 2018, AEMO published its first Integrated System Plan (ISP). The ISP initiative came out of the 2017 Independent Review into the future security of the NEM by Australia's Chief Scientist Alan Finkel (the Finkel Review), which recommended greater use of strategic planning within Australia's energy system (Finkel, 2017). The ISP has a 20-year planning horizon, is updated every two years, and it is based on detailed engineering and economic modelling of Australia's electricity network. The ISP recognises the significant changes underway in Australia, with a flattening in electricity demand from the grid and a significant shift in consumer preferences and behaviours (AEMO, 2018, p. 3). The 2020 ISP is described *as* "an actionable roadmap... to optimise consumer benefits through a transition period of great complexity and uncertainty" (AEMO, 2020, p. 9). The core proposal from AEMO is for an increase in the transmission infrastructure, identifying the "crucial role of transmission" (AEMO, 2018, p. 6) in the transition:

> The transmission grid itself requires targeted augmentation to support the change in generation mix. ... strategically placed interconnectors and REZs [Renewable Energy Zones], coupled with energy storage, will be the most cost-effective way to add capacity and balance variable resources across the whole NEM. Without adequate investment in transmission infrastructure, new VRE [renewable energy] will be struggling to connect. (AEMO, 2020, p. 13)

In other words, the ISP positions the centralised grid as continuing to be important even in the face of increased distributed generation. The ISP, therefore, advocates an increased level of investment in the electricity grid transmission lines.

Since publication of the first ISP, AEMO has grown in status to become a critical node in discussions about the future of the energy sector in Australia. In late 2020, the Australian Energy Regulator published guidelines aimed at translating the ISP into action, effective from the 2022 version of the ISP onwards (AER, 2020). In other words, the ISP will now become a regulated requirement under the National Electricity Rules in Australia. In this way, the ISP has bolstered the role of AEMO as an influential node in planning and informing Australia about its possible energy futures, albeit one particular version of the future (see Case Study 5.3).

Case Study 3.3 Islands as Energy Innovation Nodes: King Island, Australia

Urban areas are often portrayed as a natural centre of energy innovation due to the concentration of finance, people, and resources, plus multiple utility infrastructures. However, social science research shows us that rural communities are also important in innovation and learning about new energy futures (Lovell et al., 2018; Naumann & Rudolph, 2020). This is perhaps particularly true for islands because island communities are edge-of-grid: energy services are typically expensive to maintain here, so there are technical reasons why island communities tend to be at the forefront of energy innovation. Island communities also often have closer social networks and cultural ties because of their isolation; this may facilitate learning. These factors can help to explain how islands may become energy innovation nodes.

King Island is positioned north-west of the island State of Tasmania (see Fig. 3.1). The island has a small population of approximately 1500 people and a strong focus on rural industries of farming and fishing, as well as tourism. King Island's electricity grid does not have an undersea connection to Tasmania or the State of Victoria on the mainland: it is an isolated grid. Electricity is provided by a mix of renewable energy (solar, wind) and diesel generators. Diesel is imported to King Island by boat, and the whole island is powered by a 6-megawatt diesel power station.

From 2010 to 2013, the utilities on King Island undertook a range of smart grid energy innovations and upgrades, funded by the Australian Renewable Energy Agency (ARENA). The smart grid project was called KIREIP (King Island Renewable Energy Integration Project). It comprised a number of technologies and initiatives, including new solar and wind generation, a battery, flywheel, dynamic resistor and a customer demand response system (Hydro Tasmania, n.d.). The objective of KIREIP was to reduce diesel use on King Island and thereby enable the island to be more self-sufficient in energy resources. KIREIP successfully enabled a 65% reduction in diesel consumption on King Island through an entirely automated system.

KIREIP has been a notable success in terms of its replication in other places, and in this way, King Island has acted as a key node for energy innovation. The smart grid technology trialled on King Island has since been implemented (albeit in a slightly modified form) on Flinders Island

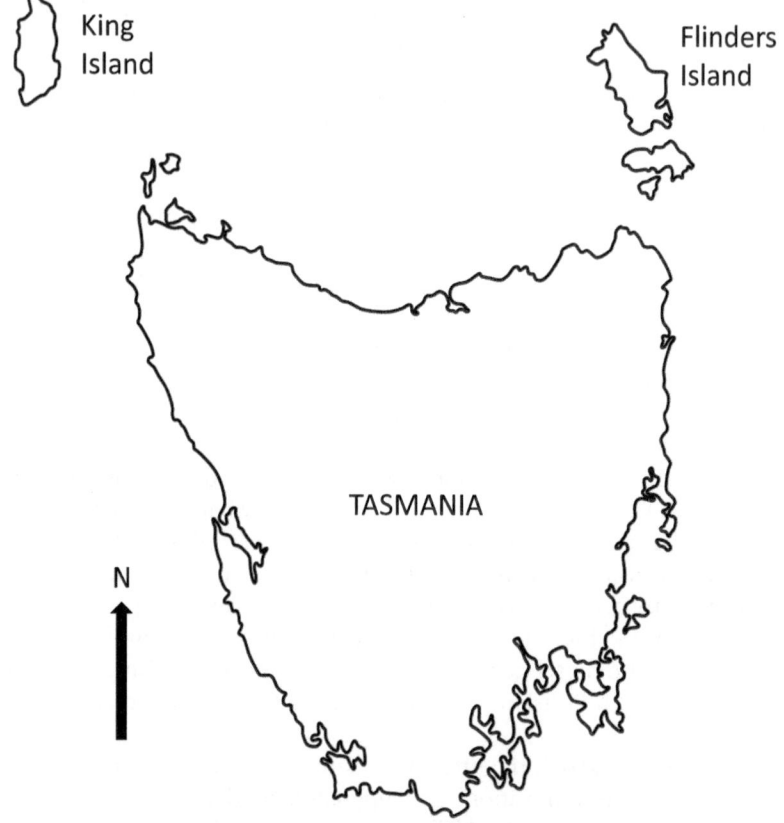

Fig. 3.1 Map of Tasmania, Australia, showing the location of King Island and Flinders Island. (Source: Original image from iStock, modified by the author)

in Tasmania, Rottnest Island in Western Australia and in Coober Pedy, a remote town in South Australia (ARENA, 2020). Hydro Tasmania has done this implementation work in conjunction with its commercial subsidiary Entura (see Entura, 2020b). After KIREIP, Entura packaged the smart grid technologies used on King Island and developed a modularized product housed within shipping containers. This Hybrid Energy Hub was implemented on Flinders Island (ARENA, 2017), a nearby island off the north-east coast of Tasmania, as the aerial photograph below shows (Fig. 3.2).

Fig. 3.2 Aerial view of the smart grid system batteries on Flinders Island, Australia. (Source: ARENA, see https://arena.gov.au/blog/flinders-island/)

The Hybrid Energy Hub was marketed on the basis of its successful implementation and performance in a rural context, on islands. In other words, the rural islandness of King Island has been important in establishing credibility for the technologies trialled on King Island and for the learning and innovation processes more generally. In a press release from ARENA announcing the start of the project on Flinders Island in 2015, the links between the two projects are highlighted:

> The Flinders Island project will build on the success of a similar project Hydro Tasmania developed on King Island… which is delivering 100 per cent renewable energy to the island. (ARENA, 2015)

And in local media coverage, the utility manager highlights the benefits of the new system on Flinders, based on the experience on King Island:

> The technology [being implemented on Flinders Island] was developed on nearby King Island, which was the first remote system capable of supplying the power needs of an entire community solely through wind and solar

energy... based on the King Island results, Flinders Island's power supply [will] become significantly more reliable. (Shine, 2017)

Entura voices similar sentiments in describing the system they implemented in a remote community in South Australia:

> The Coober Pedy hybrid renewables project builds on the King Island Renewable Energy Integration Project (KIREIP), which led the world when it first achieved 100% renewable operation using variable wind energy in 2012. (Entura, 2020a)

So, King Island acted as a node— a location where new sociotechnical energy innovations have been tested out and have then been replicated elsewhere. However, there have been some modifications to the King Island smart grid product as it has moved from place to place. For example, the customer load smart grid system, implemented as part of KIREIP, was not replicated on Flinders Island because it was found not to be frequently used on King Island and was expensive to implement. Also, some changes were made to how the technologies were packaged, as the Hydro Tasmania hybrid energy solutions manager explained:

> Hydro Tasmania took a different approach on Flinders Island in the way the system was deployed. We have modularised the enablers and we have used the platform of shipping containers. It is an approach we can deploy to other parts of the world. The Flinders Island Hub is becoming a showcase of the technology. (Shine, 2017)

So there has not been a straightforward replication of King Island smart grid technologies to different island contexts. Still, there has been significant knowledge exchange and dissemination from the King Island node.

Learning from Smart Grid Nodes

Nodes in smart grids play an important role in providing stability (keeping things the same), as well as innovating. As seen in the case studies presented in this chapter, nodes can be technologies (the digital electricity meter), places (islands), or organisations and individuals (an energy market organisation). In the table below, I summarise the key learnings from these smart grid node case studies and suggest how they might guide future practice.

Key learning	Recommendation for energy practitioners
Nodes typically have what is termed interpretative flexibility, that is, they are understood differently by different actors, and this is generally seen to be a strength; the flexibility allows them to function.	A good example of this flexibility is different understandings of electricity meters. For government and industry practitioners, meters are primarily technical nodes at the intersection of the household and utilities. However, for households, meters raise social issues about trust and equity. Appreciating these different perspectives about a node helps to plan policies and interventions and to better anticipate any problems that might arise.
Attempts are often made to replicate successful nodes elsewhere, in different contexts, but this does not always work because the things and people they are co-ordinating are subtly different.	Studies of energy innovation on islands suggest that energy innovations are usually modified along the way. Further, diversity in types of energy systems might be the new normal in the future, as communities wish to have solutions tailored to their local context, and decentralised technologies increasingly allow for this. Ideally, energy innovations are tailored to the local context in which you are working.
When nodes are positioned at the intersection of different networks (e.g., policy networks), they are particularly active and influential.	Seek to identify nodes at the intersection of different networks and notice whether they are growing in importance or waning.

REFERENCES

AEMO. (2018). *2018 Integrated System Plan for the National Electricity Market*. Australian Energy Market Operator (AEMO). Retrieved June 16, 2020, from https://aemo.com.au/-/media/files/electricity/nem/planning_and_forecasting/isp/2018/integrated-system-plan-2018_final.pdf

AEMO. (2020). *2020 Integrated System Plan for the National Electricity Market*. Australian Energy Market Operator (AEMO). Retrieved January 19, 2021, from https://www.aemo.com.au/-/media/files/major-publications/isp/2020/final-2020-integrated-system-plan.pdf

AER. (2020). *Guidelines to make the Integrated System Plan actionable*. Australian Energy Regulator (AER). Retrieved March 12, 2021, from https://www.aer.gov.au/networks-pipelines/guidelines-schemes-models-reviews/guidelines-to-make-the-integrated-system-plan-actionable

Ambrose, J. (2019, September 17). Smart energy meter rollout deadline pushed back to 2024. *The Guardian*. https://www.theguardian.com/environment/2019/sep/16/smart-energy-meter-rollout-uk-deadline-pushed-back-2024

ARENA. (2015). *Flinders Island next off-grid community to go renewable.* Retrieved December 11, 2020, from https://arena.gov.au/news/flinders-island-next-off-grid-community-to-go-renewable/

ARENA. (2017). *Flicking the switch: (Hybrid) energy comes to Flinders Island.* Retrieved February 10, 2020, from https://arena.gov.au/blog/flinders-island/

ARENA. (2020). *King Island renewable energy integration project.* Retrieved February 10, 2020, from https://arena.gov.au/projects/king-island-rene wable-energy-integration-project/

Bartlett, D., & Dibben, P. (2002). Public sector innovation and entrepreneurship: Case studies from local government. *Local Government Studies, 28*(4), 107–121.

Borgatti, S. P., Mehra, A., Brass, D. J., & Labianca, G. (2009). Network analysis in the social sciences. *Science, 323*(5916), 892–895.

Callon, M. (1986). Some elements in a sociology of translation: Domestication of the scallops and fishermen of St. Brieuc Bay. In J. Law (Ed.), *Power, action, belief* (pp. 196–233). Routledge and Kegan Paul.

Collins Dictionary. (2021). *Node – definition.* Retrieved January 4, 2021, from https://www.collinsdictionary.com/dictionary/english/node

de Wit, O., van de Ende, J., Schot, J. W., & van Oost, E. (2002). Innovation junctions: Office technologies in the Netherlands, 1880–1980. *Technology and Culture, 43*(1), 50–72.

DECC. (2012). *Smart metering implementation programme: Programme update April 2012.* Department of Energy and Climate Change. Retrieved June 13, 2019, from https://webarchive.nationalarchives.gov.uk/+/http://www.decc. gov.uk/en/content/cms/consultations/cons_smip/cons_smip.aspx#data

Entura. (2020a). *Coober Pedy hybrid renewables project.* Retrieved February 21, 2020, from https://www.entura.com.au/projects/8165/

Entura. (2020b). *Entura – Our projects.* Retrieved December 10, 2010, from https://www.entura.com.au/projects/

Etzkowitz, H., & Gulbrandsen, M. (1999). Public entrepreneur: The trajectory of United States science, technology and industrial policy. *Science and Public Policy, 26*(1), 53–62.

Finkel, A. (2017). *Independent review into the future security of the National Electricity Market: Blueprint for the future.* Commonwealth of Australia.

Google Definitions. (2021). *Definition – Node.* Retrieved January 4, 2021, from https://www.google.com/search?sxsrf=ALeKk02ubtMk9RjQFCHJUq0jXQi hfw2Uxw%3A1615424873881&ei=aW1JYLS0NaCZ4-EPyrOYqA0&q= google+dictionary&oq=google+dictionary&gs_lcp=Cgdnd3Mt d2l6EAMyCAgAELEDEIMBMggIABCxAxCDATICCAAyAggAMg IIADICCAAyAggAMgIIADICCAAyAggAOgQIIxAnOgsIABCxAxCD ARCRAjoFCAAQkQI6CggAELEDEIMBEEM6BAgAEEM6CggA EIcCELEDEBQ6DQgAEIcCELEDEIMBEBQ6BAgAEAM6BQ guELEDOgoIABCxAxCDARAKOgcIABCHAhAUOgQIABAKUP9jWOt-

YMCEAWgDcAJ4AIAB4AGIAb0dkgEGMC4xOC4zmAEAoAEBqgEH
Z3dzLXdpesABAQ&sclient=gws-wiz&ved=0ahUKEwj0jJ
POhqfvAhWgzDgGHcoZBtUQ4dUDCA0&uact=5#dobs=node

GSGF. (2012). *Global Smart Grid Federation 2012 report*. Global Smart Grid Federation (GSGF). Retrieved September 13, 2019, from https://www.smart-grid.gov/files/Global_Smart_Grid_Federation_Report.pdf

Hawkey, D., & Webb, J. (2014). *Wyndford estate district heating: Draft case study for Scottish Futures Trust's district heating delivery structures guidance*. University of Edinburgh. Retrieved March 12, 2021, from https://heatandthecity.org.uk/resources/page/2/?pid=65#resources

Hawkey, D., Webb, J., Lovell, H., McCrone, D., Tingey, M., & Winskel, M. (2016). *Sustainable urban energy policy*. Routledge.

Hydro Tasmania. (n.d.). *King Island renewable energy integration project (KIREIP)*. Retrieved July 13, 2019, from https://www.hydro.com.au/docs/default-source/clean-energy/hybrid-energy-solutions/king_island.pdf?sfvrsn=f3ad4828_2

ISGAN. (2014). *AMI case book version 2.0: Spotlight on advanced metering infrastructure*. International Smart Grid Action Network (ISGAN).

Kingdon, J. W. (2003). *Agendas, alternatives and public policies* (2nd ed.). Harper Collins College Publishers.

Lovell, H., Pullinger, M., & Webb, J. (2017). How do meters mediate? Energy meters, boundary objects and household transitions in Australia and the United Kingdom. *Energy Research & Social Science, 34*, 252–259.

Lovell, H., Hann, V., & Watson, P. (2018). Rural laboratories and experiment at the fringes: A case study of a smart grid on Bruny Island, Australia. *Energy Research and Social Science, 36*, 146–155.

Meyer, M. (2010). The rise of the knowledge broker. *Science Communication, 32*(1), 118–127.

National Grid ESO. (2020). *World's leading power system operators launch global consortium*. Retrieved December 13, 2020, from https://www.national-grideso.com/news/worlds-leading-power-system-operators-launch-global-consortium

Naumann, M., & Rudolph, D. (2020). Conceptualizing rural energy transitions: Energizing rural studies, ruralizing energy research. *Journal of Rural Studies, 73*, 97–104.

Sassen, S. (2002). *Global networks, linked cities*. Psychology Press.

Scottish Government Statistics. (2011). *Wyndford-01*. Retrieved March 13, 2021, from https://statistics.gov.scot/atlas/resource?uri=http%3A%2F%2Fstatistics.gov.scot%2Fid%2Fstatistical-geography%2FS01010358

Shine, R. (2017). *Flinders Island going for green with renewable energy hub, farewells dirty diesel*. Retrieved February 21, 2020, from https://www.abc.net.au/news/2017-11-05/flinders-island-farewells-dirty-diesel-with-renewable-energy-hub/9117108

Smart Energy GB. (2013). *Smart meter central delivery body: Engagement plan for smart meter roll-out.* Retrieved April 12, 2020, from https://www.smart-energygb.org

Star, S. L., & Griesemer, J. R. (1989). Institutional ecology, translations and boundary objects: Amateurs and professionals in Berkeley's Museum of Vertebrate Zoology, 1907–39. *Social Studies of Science, 19*(3), 387–420.

UK BIES. (2020). *Smart meter statistics in Great Britain: Quarterly report to end December 2020.* UK Department of Business, Energy and Industrial Strategy (BEIS). Retrieved January 16, 2021, from https://assets.publishing.service.gov.uk/government/uploads/system/uploads/attachment_data/file/968356/Q4_2020_Smart_Meters_Statistics_Reportv2.pdf

CHAPTER 4

Narratives

What Is a Narrative, and What Types of Narrative Do We Find in Smart Grids?

A narrative is a story about events that are connected; it is a particular way of explaining or understanding events (Cambridge Dictionary, 2021). A narrative is always a selective account: within a narrative, some things are left out, and others included. One person might see different connections between events than someone else, so the story they tell is slightly different. Hence, there are always multiple narratives about any situation, including smart grid policy programs and energy innovation more broadly. The terms narrative and story are often used interchangeably. Technically, story is more specific, as narratives are broader, with more open-ended ideas—the overall design. In contrast, stories are about the events, what happened to whom and in what order. The plot is another important variable. There is extensive discussion of narrative, story and plot within literary and cultural theory (see e.g., Herman et al., 2010; Ryan, 2017). In this chapter, I use both terms—narrative and story—but it is worth bearing in mind that there are differences between them.

Narratives demand attention because they affect how specific policies are remembered (see Chap. 5). They can also be quite subtle, emerging and coming to dominate without much attention to their origins and the reasons why that particular narrative is popular, and which others have had less attention. The reason why one energy innovation story of a policy

H. Lovell, *Understanding Energy Innovation*,
https://doi.org/10.1007/978-981-16-6253-9_4

might come to dominate and have more traction than others is to do with the specifics of what happened but also a host of external factors. These external factors include what has happened previously and peoples' memories of it, the timing of the policy and the stage in the policy process, which political party (or other organisation) is pushing for the policy change to occur, and so on. Every story about a policy is worth examining broadly to assess the context in which it has developed and to better understand why that story is the one being retold or, conversely, why it is being ignored. We see this happening in the State of Victoria's (Australia) Advanced Metering Infrastructure program (AMI) (Case Study 4.2 below), where in the domestic context, the policy failure story dominated, while internationally—in a very different context—the success story gained traction.

Characteristics of Narratives and Their Relevance to Smart Grids

A key characteristic of any popular narrative is a strong plot that does not go down too many side paths; otherwise listeners or readers get distracted and quickly lose interest. With regard to innovation, this means that stories of failure and success are amplified versions of failure or success. In other words, stories of failure do not usually have any discussion about the things that worked well, and stories of success do not mention problems, as in both cases, these would be distractions from the plot. We see this clearly in the AMI Program Case Study 4.2, below, where the dominant story of failure which took hold did not mention any successes. But there were, in fact, several successes, such as the high digital metering implementation rate.

Another characteristic of narratives to consider, which is especially relevant to better understanding energy innovation processes, is that narratives have different geographies. Some stories have strong traction and travel widely, perhaps internationally, whereas others are familiar only to a more contained (often local) group of people. Research shows us that stories about negative policies do not tend to travel far, whereas success stories do (see Lovell, 2017). The reasons for this are not hard to guess. No one likes to air their failures in public, so briefing notes on unsuccessful policies, conference talks about policy collapses and bad news media releases simply do not happen. If a policy does not work out quite as planned, the understandable tendency is not to draw attention to it. In contrast, successful policies are often promoted in media releases, through feature case studies, among networks of experts nationally and internationally, and in official reports. In the case of smart grids, we see this happening in publications by international organisations such as the International Smart Grid Action Network, publications that almost exclusively showcase successful programs (ISGAN, 2014, 2019).

The problem with stories of policy failure only being circulated locally is that there is a missed opportunity for learning. Evaluating things that went wrong and examining the reasons for this is often more productive than trying to emulate successes. This is because reflections on failures tend to provoke deeper forms of learning, such as a change in the framing of the problem and shifts in guiding values and beliefs.

Different Ways of Thinking About Narratives

The discipline areas of English and the Humanities are the natural home of narrative research, where topics range from the methodology of narrative analysis to analysis of different genres of narrative (Andrews et al., 2013; Hyvärinen, 2015). Other disciplines with a clear interest in narratives include social history, communication studies and social linguistics. Multiple definitions of narrative originate from these different areas of research; as the sociologist Riessman (2008) explores, the term narrative is used in different ways across different disciplines and has many meanings. Narratives can comprise spoken (oral), written or visual material, and even within written narratives, there are many different genres: biographies, novels, reports and so on.

Research on narratives has become increasingly popular in political science, and a methodological approach called the narrative policy framework has developed. This approach to researching policy narratives comprises three different levels of analysis—micro, meso, and macro—and examines four key elements of narratives: setting, characters, plot, and morals. Policy narratives are defined as " strategic constructions of a policy reality promoted by policy actors that are seeking to win (or not lose) in public policy battles" (Jones et al., 2014, p. 9). More generally, within political science, narratives are analysed as strategies used by policy actors to persuade others about a particular course of action. The effect of any policy narrative will vary considerably depending on the audience and context (Cairney, 2019; Fischer & Forester, 1993; Hajer, 1995).

Another relevant area of narrative research is from science and technology studies, a branch of sociology. Research here has concentrated mainly on narratives about the future, including narratives about energy futures. For example, Jasanoff and Kim (2009) developed the concept of the sociotechnical imaginary based on their comparative study of nuclear power in the USA and South Korea. They define sociotechnical imaginary as "collectively imagined forms of social life and social order reflected in the design and fulfillment of nation-specific scientific and/or technological projects" (ibid., p. 120). This narrative is about underlying visions of ideal social life, embodied in infrastructure, including smart city infrastructure (see Sadowski & Bendor, 2019). There is overlap here with research into science fiction and the way narratives of modernity and alternative versions of the future are explored through the genre of science fiction (Raven, 2017). A growing interest in energy sector narrative research is demonstrated by the 2017 publication of a dedicated special issue of the Energy Research and Social Science journal on *Narratives and storytelling in energy and climate change research*. The special issue grouped over thirty papers under the categories of stories as data, stories as inquiry and stories as process (Moezzi et al., 2017), illustrating the diverse ways in which narratives are being used to better understand energy sector innovation. A paper in this special issue, on policy narratives in the USA about smart grid interoperability standards, explains:

> Crucial for this research is the notion in discourse and narrative approaches that language does not simply mirror the world but it acts to encourage certain ways of thinking and silencing others: policy sets out a dominant conceptualization of the problem which sets limits on what can be said and felt about it. (Muto, 2017, p. 112)

Case Study 4.1 The Willing Prosumer Narrative: Householders and Their Willingness to Participate in Smart Grids

An important distinction between smart grids and previous household energy innovations is that most new smart grid technologies involve a two-way interaction between households and the electricity grid, that is, households are not just consuming electricity but producing it too. A common term for this new role for the household is *prosumer* (i.e., *pro*ducer + con*sumer* of electricity). With a range of new energy technologies now available to households, such as battery storage, electric vehicles, and rooftop solar, a narrative has emerged about the pleasure these technologies bring to households and how households universally embrace the technologies. In this narrative, a smart energy household is a happy, productive and efficient household with time and money to spare and a strong interest in being an active member of a smart grid (see Strengers, 2013 for an excellent description of the closely related 'smart utopia' narrative). Below I briefly summarise this *willing prosumer* narrative and question its origins, including the data on which it is based. The main point I wish to make is that the narrative is not based on much evidence. What is emerging from research is that households have varied, diverse responses to new smart grid technologies, which are mostly less positive than the willing prosumer narrative suggests. There are many industry and government studies about households and smart grid technologies. However, these studies are mostly techno-economic, that is, they focus on the technical feasibility and market appeal of having lots of prosumers connected to the grid (see e.g., Marchment Hill Consulting, 2012). There is little published research that has examined social factors related to prosumers, in particular, how households actually behave in their home with smart grid technologies, including how this changes over time (for exceptions see Capova et al., 2015; Watson et al., 2019). Much of the existing research is based on trials with early adopters—often time-rich technology enthusiasts—who are unlikely to represent the wider population. The International Smart Grid Action Network concludes that, despite smart grid pilots trying to support householder participation, "a consistent and integrated view on how to incentivize end users to change their behavior is still lacking" (ISGAN, 2017, p. 1). And, while there is longstanding social research on energy efficiency, photovoltaics, household demand, energy side management, and feedback more broadly (Boardman, 1994; Darby, 2006; Wade & Leach, 2003), this research is in most cases a bit different because it is not about households actively participating in smart grids, as prosumers.

Smart grid willing prosumer studies tend to use existing quantitative data about trends in household uptake to project and model into the future (CSIRO & Energy Networks Australia, 2017; Fleming et al., 2016). Theirs is an anticipatory approach, identifying trajectories and universally projecting increased future numbers of prosumers: for example, this projection from Australia's lead science agency and peak energy industry association:

scenario based modelling… identifies the possibility that up to 45% of Australia's electricity supply could be provided by millions of distributed, privately owned generators in 2050. (CSIRO & Energy Networks Australia, 2017, p. 2)

There is an underlying assumption here that households will be willing players in new energy technologies and network sharing. In other words, the studies that promote the willing prosumer narrative are mostly based on assumptions about likely positive household responses rather than actual evidence. For example, consider these statements from the European Commission and then the Council of European Energy Regulators (CEER):

Smart grids enable new market actors, such as aggregators and energy service companies, to offer new types of services to consumers, allowing them to adjust their consumption and reap the benefits of flexibility provided to the grid. (European Commission, 2021b)

The emergence of smart technologies is driving change in energy markets. It is beginning to change the traditional role of the customer, providing them with greater opportunities. (CEER, 2018, p. 7)

What is mostly left out of the willing prosumer narrative is an appreciation of the diversity and complexity of household responses to smart grids. The narrative is based on an ideal type of household. It is a household that remains engaged, and their willingness to participate does not tail off over time, after the novelty of the new energy technology fades.

The context that has allowed the willing prosumer narrative to flourish can be understood by looking at industry and government motivations and interests. On the commercial side, the willing prosumer narrative has been driven by businesses wishing to expand their interests into the growth area of smart grids and the digitalisation of utility infrastructure more broadly. These include large international companies such as IBM and Cisco (see Sadowski & Bendor, 2019), and is not surprising given that investment in smart grids globally per year is as much as US$275 billion (IEA, 2020). For some smart grid companies, the market is direct to households but for most, their customers are other organisations in the energy sector. So, these companies have a financial interest in promoting the willing prosumer narrative to other businesses and governments. The household is positioned within the narrative as, first and foremost, a compliant household, a household that uses their new smart grid product or technology efficiently to effectively manage their own electricity and on behalf of the grid. For example, the Director of Sustainable Energy at Cisco, an international internet and digital business, explains the role of consumers like households within smart grids as follows:

> A Smart Grid will enable consumers to manage their own energy consumption through dashboards and electronic energy advisories. More accurate and timely information on electricity pricing will encourage consumers to adopt load-shedding and load-shifting solutions that actively monitor and control energy consumed by appliances. (Frye, 2008, p. 7)

And a journalist in the popular trade magazine RenewEconomy similarly states:

> many customers have begun taking more direct control of the cost, reliability, and green mix of their energy supply. They are enabled on this journey by a convergence of new, widely available technologies that can automate and fully monetise their energy resources. (Mouat, 2016)

Governments too are drivers of the willing prosumer narrative, investing increasing amounts of resources in smart grids. Governments have promoted smart grids as a core element of the new digital economy (see e.g., DECC & Ofgem, 2014; and European Commission, 2021a). There are several pressing policy problems that smart grids have the potential to solve, including rising electricity prices and decarbonisation of the grid, so governments also have an interest in promoting smart grids for these reasons (Mission Innovation, 2019; UNECE, 2015). As the science and technology studies scholar Sachiko Muto (2017) explains, there is a *hero technology* narrative with smart grids, in which smart grids provide a solution to a host of policy problems. In the willing prosumer narrative, the household is the key actor pivotal to the smart grid hero technology working.

CASE STUDY 4.2 THE NARRATIVE OF POLICY FAILURE
IN THE STATE OF VICTORIA'S DIGITAL METERING PROGRAM

Digital meters came onto the mass market between 2005 and 2010, offering many more features than the traditional spinning disc accumulation meter (see Case Study 3.1). Countries around the world have struggled to work out the best way to implement these next generation meters in households and small businesses. The two main implementation options that have been used, to varying degrees of success, are mandatory (government-initiated) and voluntary (customer opt-in/customer choice). There are advantages and disadvantages of each approach. An advantage of a mandatory approach is the efficiency of implementation, as a whole street or neighbourhood can have new meters installed at the same time, reducing travel time and costs. Another plus is the benefit of having real-time customer data from the entire network (so-called digital meter saturation). But, as with all large-scale government infrastructure programs, costs can quickly escalate and promised financial benefits have often not materialised.

Australia is an interesting example internationally because it has done both types of implementation. The State of Victoria adopted a mandatory approach under its Advanced Metering Infrastructure (AMI) program (2009 to 2013). Initially, all the other states were to follow, but, in the end, the Australian National Electricity Market implemented a voluntary customer-led program. Households are not obliged to accept a digital meter unless they are moving house, their existing old-style meter is faulty, or they opt to change tariff. This approach is often referred to as new and replacement digital metering implementation, and in effect, mixes voluntary and mandatory elements. Why was there was a switch from a mandatory to a voluntary mode of implementation in Australia? The decision stems from the strong narrative of failure that emerged from the State of Victoria's AMI program (see also Case Study 2.3).

In the late 1990s, Victoria became the first state in Australia to privatise its electricity sector. Partly because of this, it was keen to go ahead with digital metering so that its newly privatised market could function better, as digital meters allow greater choice of tariffs and easier switching of electricity company provider. In 2004 state government approval was given in Victoria to proceed with an interval metering program. Interval meters are basic digital meters that collect consumption data in a digital form but do not transmit or communicate the data remotely. As technology choice rapidly improved in the mid-2000s, the approval in Victoria was changed to advanced meters in 2006. Advanced meters have communications embedded and so can transmit data remotely, without having to be manually read. The AMI program ran from 2009 to 2013 and resulted in 2.3 million digital meters being installed in 93% of homes and small businesses in Victoria.

On the face of it, Victoria's AMI program sounds like a wonderful success story with a really high implementation rate and delivery of modern utility infrastructure across the whole of the state. This story or narrative is certainly one that could be told about the AMI program. In my research, I found that there were many stories about the AMI program and that stories were a significant feature of how people sought to understand what happened and how to move forward.

The local story of failure: This is the dominant narrative that emerged about the Victorian AMI program. The program was a failure from start to finish: it was expensive, households and small businesses bore the costs, the government and utilities mismanaged it, the government did not take full responsibility for the program, the advantages of digital metering were overstated and the disadvantages overlooked. These criticisms were widely reported in the media and also through official reviews. For example, the Victorian Auditor-General reviewed the program twice (2009, 2015), and both times it was heavily criticised. Under this narrative, there is no room for successes. Although there was a high implementation rate (93%), the program was costly. Part way through the program, the state government ruled out some planned benefits of the meters in terms of tariff changes (to time-of-use tariffs). Utilities were unhappy about this, as this was a key part of their business case for investment. In this story, the AMI program was a dismal failure that meant there was no possibility of other Australian states implementing mandatory digital metering programs, as had been initially planned. Digital metering had become politically sensitive.

The international story of success: In stark contrast to the story of failure outlined above, there is an alternative story about how successful the AMI program was. This story was harder to locate, but it became apparent to me while undertaking interviews with key AMI program decision-makers in the Victorian government. There were several passing references to the high numbers of international delegations coming to Victoria to learn about the AMI program. It transpired that, at least in some countries, the story of failure was not the dominant story about the AMI program. Instead, international governments and utilities were eager to find out how Victoria had managed to achieve such success in its digital metering implementation. Many international delegations visited Victoria to hear about and learn from the AMI program, so they could repeat its success back at home, as one interviewee in the Victorian government commented: "last week we had a group from Malaysia and you know they were really engaged... and interested in our experience...So we get a lot of people coming to see what we've done" (Interview, September 2015). I also witnessed this first-hand on a research visit to London in 2019, when the organisation overseeing the implementation of the UK's voluntary digital metering program wanted to meet with me to discuss Victoria's AMI program. It was considering transitioning to a mandatory digital metering program and saw what had been done in Victoria as highly successful and as a possible model for the UK.

These contrasting stories of success and failure are the two main stories in circulation. But there are other ones too. For example, there is a story of *wasted learning* about how there was lots of useful learning from Victoria, but this learning could not be applied because of the political reaction to the things that went wrong and the bad press around them. Because of the subsequent decision to go ahead with a voluntary program in other states, the AMI program was not able to be learnt from. If mandatory metering implementation had progressed as planned in other states, it could have been much better and run much more smoothly than in Victoria. In other words, positive changes could have been made based on what happened in Victoria, as a general manager at a Victorian distribution utility described:

> on our calculation they [the other Australian States] could probably rollout the program for 30 to 40 percent of the cost Victoria did because we've learnt all the lessons and all the technology is off the shelf now... we have actually had all the problems and solved them. (Interview, May 2017)

Another story is about *rapid technology change and the timing of policy interventions*. This tale is of how the Victorian government moved too early, at a time when digital meters were only evolving rapidly as a technology, and were not well tested. Many of the problems that arose in Victoria were due to the early type of digital meters that were implemented. There was no mass production of advanced (communications-enabled) digital meters when the Victorian government committed to the AMI program in 2006. Victoria gave digital meters a bad reputation, but metering technology has considerably improved since then. A sub-plot of this story is about the communications technology used to support the meters in Victoria. Most utilities decided to go with one type of communications technology (mesh), but one utility—AusNet—opted to try something different (WiMax). Despite repeated attempts to encourage the WiMax communications technology to work, and with much money spent, AusNet eventually admitted defeat and switched to the mesh technology used by the other utilities.

CASE STUDY 4.3 NARRATIVES THAT COMPETE WITH SMART GRIDS: THE HYDROGEN ECONOMY AND OFF-GRID

The policy narrative of smart grids has declined in popularity in recent years. Its heyday was around 2010, when there were lots of smart grid initiatives world-wide, with government funding, trials, standards development and so on. The graph below (Fig. 4.1), showing Google searches for "smart grid", illustrates this drop-off in interest.

More specifically, it is possible to see this trend in the way Australian smart grid initiatives have been referenced within international policy documents, with a peak in 2012 and a decline since then (see Fig. 2.2, Chap. 2). Like most other things in our lives, there are fashion cycles with policies, and ideas fall in and out of favour. Smart grids are no different. It is often that a few things do not go as well as hoped during implementation, and then it suddenly begin to look like a less exciting policy option. As an energy consultant and a regulator explained to me:

> Why are you doing a research project on smart grids? No-one talks about smart grids any more. If I were you I would research something else. (Interview, Energy Consultant, April 2015)

> We don't talk about smart grids at all now really….it all seemed a little bit gimmicky, it seemed like a marketing idea rather than a wholesale change in mindset. (Interview, Californian Regulator, March 2016)

Research on policy fashions seeks to distinguish between the narrative and what is happening on the ground, that is, between rhetoric and practice (also Naim, 2000; see for example Peck & Theodore, 2015 who analyse the speed at which new policy fashions circulate internationally; Pollitt et al., 2001). There is mixed evidence about whether the implementation of smart grids still continues as before—and it is just the policy narrative that has tailed off and become less popular—or whether there has been a wholesale shift away from smart grids, both in rhetoric and practice (Lovell, 2019).

Usually as part of the policy cycle—during the waning phase—a new policy idea comes along with a shiny new narrative that has strong appeal. In Australia, two such narratives that have taken some of the attention

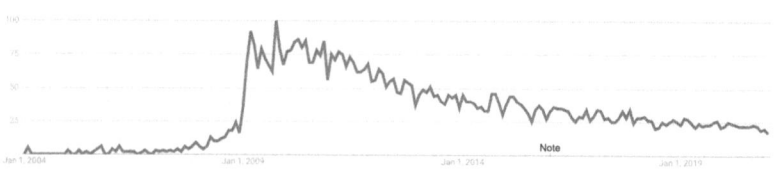

Fig. 4.1 The changing level of interest in smart grids over time. (Source: Google Trends)

previously directed to smart grids are the hydrogen economy and off-grid initiatives. The rise of these two competing narratives in Australia is reflective of wider global trends. For example, in relation to hydrogen, in the last few years, the industry-led international Hydrogen Council was established (2017), the first World Hydrogen Congress took place (2020), and the International Energy Agency produced a landmark report, *The Future of Hydrogen* (IEA, 2019).

The Hydrogen Economy

There is growing interest internationally in using hydrogen as an energy fuel, a narrative and policy agenda that to some extent competes with smart grids. Hydrogen can be produced from water with electricity using electrolysis. This process is very energy-intensive, but the technology has been tried and tested over several decades and is improving. The interest in hydrogen as a fuel is in large part driven by the problem of climate change, as well as local air pollution. When hydrogen is burned, the only waste product is water. However, we must also think about how hydrogen is produced in the first place: if it is made using electricity produced by fossil fuels, then it still contributes to climate change. There are lots of aspects of the hydrogen economy still to be worked out. However, some countries are starting to invest significantly in hydrogen and with long term plans. Japan and South Korea, for instance, wish to mass import hydrogen produced using renewable electricity, including from Australia (Japanese Hydrogen and Fuel Cell Strategy Council, 2019; South Korean Government, 2019).

The hydrogen economy narrative stands in strong contrast to the smart grids narrative. There is no smart digitalisation, network efficiency or information and communications technology meets utility sector expertise in this narrative. Nevertheless, it is a competitor of the smart grids narrative, because the end product—improved energy services to consumers—is the same. It is just a different way of getting there. Instead of real-time feedback and big data, the narrative of the hydrogen economy is populated by quite different things: shipping container transport, large-scale manufacturing, fertilisers, and pipelines. In many ways, it is back to the old school energy sector of large industrial sites and engineering.

There is a good example of the role of narrative in technology development in the Australian 2019 *National Hydrogen Strategy*, which commences with a quote from Jules Verne's 1874 novel *The Mysterious Island:*

> In 1874, science fiction author Jules Verne envisioned a future in which 'water will one day be employed as fuel, that hydrogen and oxygen which constitute it, used singly or together, will furnish an inexhaustible source of heat and light, of an intensity of which coal is not capable. Someday the coal-rooms of steamers and the tenders of locomotives will, instead of coal, be stored with these two condensed gases, which will burn in the furnaces with enormous calorific power.' Verne's prescient vision has inspired governments and entrepreneurs in the 145 years since. (COAG Energy Council, 2019, p. v)

It is a useful reminder of how policy decisions are guided by a mix of cultural, social, political, and economic factors. In other words, any decision about whether to invest in hydrogen in Australia, or elsewhere, is not just about economics, despite the widespread use of the term *hydrogen economy*. For Australia, it is also about its vision for the future and its position in the world. As the Council of Australian Governments states: "Australia has the resources, and the experience, to take advantage of increasing global momentum for clean hydrogen and make it our next energy export." (COAG Energy Council, 2019, p. viii)

Off-grid

Another narrative that has emerged that has taken some attention from the smart grids narrative is *off-grid*, also referred to as micro-grids or decentralised energy. This narrative is in strong contrast to the idea of a large seamless, efficient smart grid. At its heart, it is about the fragmentation and break down of the existing grid into isolated pockets of electricity generation and consumption. These pockets could be at the household (off-grid) or community (micro-grid) level. This off-grid narrative has been made possible in part by the development of new technologies as well as the refinement of existing technologies, particularly electricity storage technologies such as household-level batteries. Key terms that populate this narrative are self-sufficiency, distrust in utilities, rural and remote, resilience, and battery health. So, like the hydrogen economy narrative, it is quite a different narrative from smart grids.

In Australia, the off-grid narrative has become more prominent since the 2019/2020 bushfires. As many remote communities lost their electricity transmission lines during the bushfires, discussion has grown about the benefits of rebuilding the electricity infrastructure as isolated grids—off-grid communities—rather than reconnecting these communities to the main grid. In this way, we see how off-grid has become a more mainstream policy option. The more general situation in Australia is one where off-grid makes sense because of the *stringyness* of Australia's electricity grid. Australia's east coast has the longest interconnected transmission network in the world (ENA, n.d.), with long feeder lines supplying electricity to often just a handful of rural customers. Not surprisingly perhaps, electricity supply to remote communities from the main grid is not always reliable (see also Case Study 2.2). So, an element of the off-grid narrative in Australia is about greater security and reliability of supply.

In terms of translating the off-grid narrative into a reality, that is, the actual implementation of off-grid infrastructure, surprisingly little data is being collected (see Case Study 5.2). For off-grid households, in particular, much of this activity is off the radar of governments and utilities. The emphasis of analysis has tended to be on modelling future scenarios that use projections based on assumptions—rather than actual data—about the number of off-grid households (see e.g., Brinsmead et al., 2015; Clean Energy Council, 2015; CSIRO, 2013; Graham et al., 2015; Szatow & Moyse, 2014). So, despite the increasing use of the off-grid narrative, data collection in Australia remains centred on the existing utilities and large-scale centralised energy infrastructure.

LEARNING FROM SMART GRID NARRATIVES

Stories have played an important role in helping us to simplify and make sense of new energy innovations such as smart grids. The handful of smart grid narratives that I have presented in this chapter show how contradictory narratives co-exist and compete for our attention, how they travel across time and place, and how a successful narrative can influence the future of energy innovation. In the table below, I summarise the key learnings from these smart grid narrative case studies and suggest how they might guide future practice.

Key learning	Recommendation for energy practitioners
Narratives are useful to study not only because of the things, people, and organisations that they speak to but also because of the things that get left out of them, the gaps or silences.	It is important to notice the things excluded or unsaid within a particular narrative and to pay attention to the cohesiveness of narratives and their particular framings of the problem and its solutions.
Evidence that runs counter to a popular policy narrative tends to be ignored.	The evidence base behind a particular narrative needs to be carefully considered: some narratives become very popular because they are a good strategic fit, and organisations with vested interests are driving the narrative, but there may actually be little empirical data to substantiate the narrative (e.g., the willing prosumer narrative).
Learning from energy sector failure is more difficult than learning from success because there is much less information circulating about failures.	Publishing detailed information about things that did not work well with an energy policy or new initiative, and not only publicising the success stories, supports successful energy innovation. Publication could be several years later, without the pressure of heightened media attention, and once there is more data on the longer-term benefits and disadvantages.

REFERENCES

Andrews, M., Squire, C., & Tamboukou, M. (2013). *Doing narrative research.* Sage.

Boardman, B. (1994). Energy efficiency measures and social inequality. In M. Bhatti, J. Brooke, & M. Gibson (Eds.), *Housing and the environment: A new agenda* (pp. 107–127). Chartered Institute of Housing.

Brinsmead, T. S., Graham, P., Hayward, J., Ratnam, E. L., & Reedman, L. (2015). *Future energy storage trends: An assessment of the economic viability, potential uptake and impacts of electrical energy storage on the NEM 2015–2035.* CSIRO, Brisbane, Australia.

Cairney, P. (2019). *Policy in 500 words: The narrative policy framework.* Retrieved December 4, 2020, from https://paulcairney.wordpress.com/2019/01/28/policy-in-500-words-the-narrative-policy-framework/

Cambridge Dictionary. (2021). *Narrative – Definition.* Retrieved December 14, 2020, from https://dictionary.cambridge.org/dictionary/english/narrative

Capova, K., Powells, G., Bulkeley, H., & Lyon, S. (2015). *Domestic customers: Energy practices and flexibility.* University of Durham – Customer Led Network Revolution Report for OFGEM UK. http://www.networkrevolution.co.uk/wp-content/uploads/2015/02/CLNR-L102-Domestic-Customers-Energy-practices-and-flexibility2.pdf

CEER. (2018). *Council of European Energy Regulators Report on smart technology development.* Council of European Energy Regulators (CEER).

Clean Energy Council. (2015). *Australian Energy Storage Roadmap.* Clean Energy Council, Canberra, Australia.

COAG Energy Council. (2019). *Australia's national hydrogen strategy.* Retrieved March 4, 2020, from https://www.industry.gov.au/sites/default/files/2019-11/australias-national-hydrogen-strategy.pdf

CSIRO. (2013). *Change and Choice: The Future Grid Forum's analysis of Australia's potential electricity pathways to 2050.* CSIRO Future Grid Forum.

CSIRO, & Energy Networks Australia. (2017). *Electricity network transformation roadmap: Final report.* Retrieved August 17, 2019, from https://www.energy-networks.com.au/resources/reports/electricity-network-transformation-roadmap-final-report/

Darby, S. (2006). *The effectiveness of feedback on energy consumption* (Vol. 486). UK Government.

DECC, & Ofgem. (2014). *Smart grid vision and routemap.* Department of Energy and Climate Change (DECC) Ofgem. Retrieved October 13, 2017, from https://assets.publishing.service.gov.uk/government/uploads/system/uploads/attachment_data/file/285417/Smart_Grid_Vision_and_RoutemapFINAL.pdf

ENA. (n.d.). *Guide to Australia's Energy Networks.* Energy Networks Australia (ENA).

European Commission. (2021a). *The European digital strategy.* Retrieved February 12, 2021, from https://ec.europa.eu/digital-single-market/en/content/european-digital-strategy

European Commission. (2021b). *Smart grids and meters.* Retrieved January 18, 2021, from https://ec.europa.eu/energy/topics/markets-and-consumers/smart-grids-and-meters_en

Fischer, F., & Forester, J. (1993). *The argumentative turn in policy analysis and planning*. UCL Press Ltd.

Fleming, A., Mankad, A., Cavanagh, K., Price, C., Behrens, S., Haigh, N., & Lim, O. (2016, August). Guilty as charged? *Ecogeneration*, 2731.

Frye, W. (2008, November). *Smart grid: Transforming the electricity system to meet future demand and reduce greenhouse gas emissions*. Retrieved June 12, 2019, from https://www.cisco.com/c/dam/en_us/about/ac79/docs/wp/Smart_Grid_WP_1124aFINAL.pdf

Graham, P., Brinsmead, T., Reedman, L., Hayward, J., & Ferraro, S. (2015). *Future grid forum – 2015 refresh: Technical report*.

Hajer, M. A. (1995). *The politics of environmental discourse: Ecological modernisation and the policy process*. Clarendon Press.

Herman, D., Manfred, J., & Marie-Laure, R. (2010). *Routledge encyclopedia of narrative theory*. Routledge.

Hyvärinen, M. (2015). Analyzing narrative genres. In *The handbook of narrative analysis* (pp. 178–193).

IEA. (2019, June). *The future of hydrogen: Seizing today's opportunities*. International Energy Agency (IEA). Retrieved February 10, 2020, from https://www.iea.org/reports/the-future-of-hydrogen

IEA. (2020). *IEA tracking report – Smart grids*. Retrieved January 13, 2021, from https://www.iea.org/reports/smart-grids

ISGAN. (2014). *AMI case book version 2.0: Spotlight on advanced metering infrastructure*. International Smart Grid Action Network (ISGAN).

ISGAN. (2017). *ISGAN Annex 7 (Smart Grid Transitions) Policy Brief No.1 – Phase-sensitive Enabling of Household Engagement in Smart Grids*. International Smart Grid Action Network (ISGAN). Retrieved May 14, 2021, from https://www.iea-isgan.org/wp-content/uploads/2018/02/ISGAN_PolicyBrief_HouseholdEngagementInSmartGrids_2017.pdf

ISGAN. (2019). *AMI CASE Case05 – ITALY*. International Smart Grid Action Network (ISGAN). Retrieved September 10, 2019, from http://www.iea-isgan.org/ami-case-case05-italy/

Japanese Hydrogen and Fuel Cell Strategy Council. (2019). *The strategic road map for hydrogen and fuel cells – Industry-academia-government action plan to realize a "Hydrogen Society"*. Retrieved March 3, 2020, from https://www.meti.go.jp/english/press/2019/pdf/0312_002b.pdf

Jasanoff, S., & Kim, S.-H. (2009). Containing the atom: Sociotechnical imaginaries and nuclear power in the United States and South Korea. *Minerva*, *47*(2), 119.

Jones, M., Shanahan, E., & McBeth, M. (2014). *The science of stories: Applications of the narrative policy framework in public policy analysis*. Springer.

Lovell, H. (2017). Are policy failures mobile? An investigation of the Advanced Metering Infrastructure Program in the State of Victoria, Australia. *Environment and Planning A, 49*(2), 314–331.

Lovell, H. (2019). The promise of smart grids. *Local Environment, 24*(7), 580–594.

Marchment Hill Consulting. (2012). *Energy storage in Australia: Commercial opportunities, barriers and policy options.* Report for the Clean Energy Council, Australia.

Mission Innovation. (2019). *Smart grids innovation challenge country report 2019.* Mission Innovation. Retrieved February 4, 2020, from https://www.mi-ic1smartgrids.net/wp-content/plugins/dms/pages/file_retrieve.php?obj_id=154

Moezzi, M., Janda, K. B., & Rotmann, S. (2017). Using stories, narratives, and storytelling in energy and climate change research. *Energy Research & Social Science, 31*, 1–10.

Mouat, S. (2016). A new paradigm for utilities: The rise of the "prosumer". *RenewEconomy.* https://reneweconomy.com.au/new-paradigm-utilities-rise-prosumer-26384/

Muto, S. (2017). From laissez-faire to intervention: Analysing policy narratives on interoperability standards for the smart grid in the United States. *Energy Research & Social Science, 31*, 111–119.

Naim, M. (2000). Fads and fashion in economic reforms: Washington Consensus or Washington Confusion? *Third World Quarterly, 21*(3), 505–528.

Peck, J., & Theodore, N. (2015). *Fast policy: Experimental statecraft at the thresholds of neoliberalism.* The University of Minnesota Press.

Pollitt, C., Bathgate, K., Caulfield, J., Smullen, A., & Talbot, C. (2001). Agency fever? Analysis of an international policy fashion. *Journal of Comparative Policy Analysis, 3*(3), 271–290.

Raven, P. G. (2017). Telling tomorrows: Science fiction as an energy futures research tool. *Energy Research & Social Science, 31*, 164–169.

Riessman, C. K. (2008). *Narrative methods for the human sciences.* Sage.

Ryan, M. L. (2017). Narrative. In *A companion to critical and cultural theory* (pp. 517–530).

Sadowski, J., & Bendor, R. (2019). Selling smartness: Corporate narratives and the smart city as a sociotechnical imaginary. *Science, Technology, & Human Values, 44*(3), 540–563.

South Korean Government. (2019). *Roadmap to the worlds' best leading hydrogen economy.* South Korean Department of Energy New Industry. Retrieved July 13, 2020, from http://www.motie.go.kr/motie/ne/presse/press2/bbs/bbsView.do?bbs_cd_n=81&cate_n=1&bbs_seq_n=161262

Strengers, Y. (2013). *Smart energy technologies in everyday life: Smart utopia?* Palgrave Macmillan.

Szatow, T., & Moyse, D. (2014). *What happens when we unplug? Exploring the consumer and market implications of viable, off-grid energy supply.* Energy for the People and the Australian Technology Association (ATA), Melbourne, Australia.

UNECE. (2015). *Electricity system development: A focus on smart grids, overview of activities and players in smart grids.* United Nations Economic Commission For Europe (UNECE). Retrieved September 10, 2019, from https://unece. org/fileadmin/DAM/energy/se/pdfs/eneff/eneff_h.news/Smart.Grids. Overview.pdf

Wade, J., & Leach, M. (2003). *Energy efficiency in UK energy policy: A step change towards a low carbon economy.* ECEE 2003 Summer Study. Retrieved March 13, 2015, from http://www.ukace.org/research/eceee_carbon.pdf

Watson, P., Lovell, H., Ransan-Cooper, H., Hann, V., & Harwood, A. (2019). *Consort Bruny Island battery trial project final report – Social Science.* Report Produced for Australian Renewable Energy Agency (ARENA). https://arena. gov.au/assets/2019/06/consort-social-science.pdf

CHAPTER 5

Nostalgia

What Is Nostalgia, and What Types of Nostalgia Do We Find in Smart Grids?

The modern meaning of nostalgia is about happy memories of the past, a looking back to the past with longing and fondness, as the psychologist Constantine Sedikides explains, it is "a sentimental longing for one's past" (Sedikides et al., 2008, p. 305). It is a term normally used in a personal, individual sense to describe memories of the past and a longing for how things were. There are associations between nostalgia and homesickness. In fact, homesickness was the original medical meaning of nostalgia; from the Greek *nostos*, meaning return to home and *algos*, pain. Over time the meaning of nostalgia has changed; it has switched from a place to a time, and it no longer refers to illness. Historically the term nostalgia was always used in a negative sense—denoting difficulty living and thriving in the present moment, because of a sense of loss and a fixation on another place or the past, as the communication and media scholars Pickering and Keightley explain, nostalgia was characterised by "a defeatist attitude to present and future" (Pickering & Keightley, 2006, p. 920). But the contemporary meaning has shifted substantially, and it is now more about looking back to the past with fondness, as English literature scholar Dames eloquently describes:

H. Lovell, *Understanding Energy Innovation*,
https://doi.org/10.1007/978-981-16-6253-9_5

> Longing for the vanished past; a registration of loss that is nonetheless plea-
> surable, even an indulgence (Dames, 2010, p. 271)

Applying the idea of nostalgia to energy innovation might seem like quite a big leap, but there are connections between technologies and a longing for the past. Nostalgia can be applied to time periods or specific objects. Some people might have nostalgia for the era before the internet and widespread computing at home, a yearning for a simpler life without so much digital connectivity. Nostalgia can also be thought about at an organisational level, with organisations fixated on existing longstanding ways of doing things—looking to the past much more than to the future (Czarniawska, 1997). So nostalgia can act in opposition to innovation because of the desire for a return to older ways of doing things, as English literature scholars Atia and Davies (2010, p. 181) explain:

> Nostalgia is always suspect. To give ourselves up to longing for a different
> time or place, no matter how admirable its qualities, is always to run the risk
> of constricting our ability to act in the present.

Nostalgia is often a crucial part of narratives. You will see this in the narrative about off-grid households, discussed in the case study below (see also Case Study 4.3, Chap. 4). This narrative is a complex mix of nostalgia and innovation: nostalgia for a simpler way of life that is self-sufficient in resources but with innovation in the form of new battery technologies and electricity generation capabilities.

DIFFERENT WAYS OF THINKING ABOUT NOSTALGIA

Nostalgia is about memories, so history and cultural studies are important areas of scholarship. For instance, a special issue of the journal *Memory Studies* (2010; Volume 3, Issue 3) is dedicated to nostalgia. The issue includes articles on a diverse range of topics, from the relationship between anthropological nostalgias and indigenous self-understanding (Whitehead, 2010), to the role of nostalgia in the notion of ecological sustainability (Davies, 2010). Nostalgia spans the disciplines of history and anthropology, literary criticism and art history, environmental and cultural studies, psychology, media studies, sociology, and political science. Nostalgia is, therefore, a highly interdisciplinary area of study, as Pickering and Keightley (2006, p. 922) explain:

nostalgia has been used in many fields of study as a critical tool to interrogate the articulation of the past in the present, and in particular, to investigate sentimentally inflected mediated representations of the past

Psychological research into nostalgia draws on the origins of nostalgia as an illness, with a focus on the individual. Psychologists have found that nostalgia transcends different social groups and age categories and can be a positive experience that allows individuals to have resilience and cope with challenges (Sedikides et al., 2008). However, there are mixed views about nostalgia, and some see it as negative; as Pickering and Keightley (2006, p. 921) explain: "*Nostalgia can be both melancholic and utopian.*" Scholars draw a distinction between personal and social (or historical) nostalgia (Routledge, 2015). The emphasis of research on social nostalgia is about how memories are generated within particular communities or cultures through the lens of nostalgia:

> nostalgia is read not only as a symptomatic state of mind, but also as a way of shaping and directing historical consciousness. (Atia & Davies, 2010, p. 182)

The influence on consciousness characteristic of nostalgia can be dangerous. It is possible to develop false nostalgia for time periods that never existed in the way people remember, as the psychologist Routledge explains:

> Perceptions of the past can… be inaccurate. Time allows us to make sense of and extract meaning from the past, but this process can also lead us to romanticize it. (Routledge, 2017)

Routledge explores how false nostalgia can be used as a political tool to subtly make people feel anxious and mobilise them into the desired action, noting how "nostalgia has the power to mobilize and inspire people when they are most vulnerable" (ibid.).

There is scholarship on the technologies of memory (i.e., the devices and things that help us to remember (see Van House & Churchill, 2008)), but very little existing research into the relationship between energy technologies and nostalgia. One example is research by Hanel and Hård (2015), who use the concept of nostalgia to examine nuclear power. They show how nostalgia for heavy-water nuclear plants in Sweden and West

Germany meant they were slow to take up new light-water designs. Another example, albeit outside of the energy sector, is research by anthropologist Ray Cashman (2006) on nostalgia for old farming equipment in a community in Northern Ireland. Old farming tools such as threshers have been restored and displayed by several community members, in the process helping the community to adapt to the pace of change and to resolve religious differences. Cashman finds positive meaning in the nostalgia within this rural area:

> The amateur curators of the Derg Valley's past material culture are not infected [with] an unthinking or merely sentimental nostalgia... they quite sanely challenge both the presumption that modernization equals positive progress and the impulse to romanticize the past. (ibid., p. 148)

In other words, nostalgia can be productive and does not necessarily have to be a negative influence on technological progress and innovation:

> Nostalgia may also be seen as seeking a viable alternative to the acceleration of historical time, one that attempts a form of dialogue with the past and recognizes the value of continuities in counterpart to what is fleeting, transitory and contingent. (Pickering & Keightley, 2006, p. 923)

Nostalgia can be applied to anything. It is something felt by those in exile and migrants towards their country of origin, as well as by individuals towards a technical object. It is also increasingly recognised that nostalgia can be quite radical (i.e., conducive to innovation), depending on what it is being applied to:

> Recent historians of nostalgia have shown persuasively that nostalgia can become creative or radical by virtue of its object, by its being nostalgic for anything from farming equipment in Northern Ireland to pre-scientific English agrarian socialism or the unfulfilled promises of the East German state. (Atia & Davies, 2010, p. 183)

This framing is relevant for considering the application of nostalgia to energy innovation and smart grids. Nostalgia for traditional ways of doing things in the energy sector and within the home can actively shape the innovation process, and not always in reactionary or retrogressive ways.

CASE STUDY 5.1 MEMORIES OF INTERNATIONALLY
PIONEERING SMART GRID EXPERIMENTS AND THEIR LEGACY

A handful of pioneering smart grid experiments is well known internationally by those working in the field. These include the digital metering program in Italy (the Telegestore project), California's smart grid program, and Ontario's grid modernisation program in Canada. Figure 5.1 below shows the timeline of these programs and other key international smart grid initiatives from 1999 to 2020. Although I use the term experiments, these examples are, in fact, diverse in origin and operation, ranging from innovative policy programs in diverse geographical areas (nation states, regions/states) to more discrete experiments in individual cities and smaller locales. Most involve a mix of public and private sector organisations, usually backed by public sector funding. What unites these smart grid experiments is that internationally they are well known and have been heavily profiled and discussed within the smart grid sector. These are early examples of smart grids—the first wave—and have become part of policy narratives about smart grids internationally.

What are the memories of these internationally pioneering smart grid experiments? What is their legacy? One notable thing regarding how they are talked about, which relates to nostalgia, is that they are mostly discussed in positive terms: they are remembered with sentimentality and fondness. For instance, the early smart grid program in California is

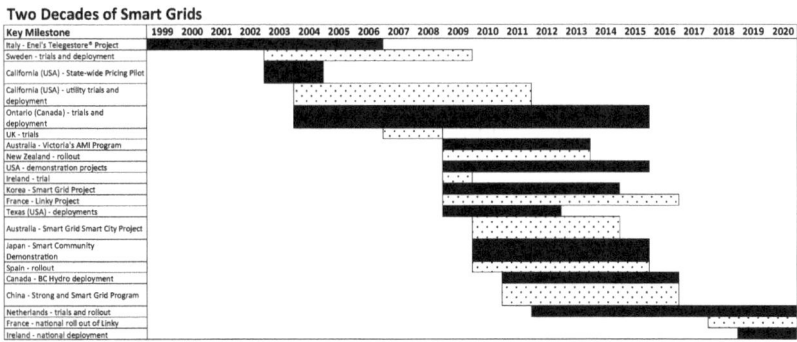

Fig. 5.1 Timeline of international smart grid projects. (From data analysis conducted by Dr Cynthia Nixon, University of Tasmania, 2020)

positioned by the International Smart Grid Action Network (ISGAN) as being a leader and outstanding:

> SDG&E [the San Diego Gas and Electric Company] stands out as having an outstanding AMI outreach and deployment methodology…. One of the early implementers, SDG&E… [were] the first utility in the USA to cover their entire service territory with gas and electric smart meters…SDG&E also did two years of deep design work prior to doing any deployment. Customers were even involved in a co-design process prior to the first AMI deployment in 2009….Unsurprisingly, in California and the broader USA, SDG&E is known as a leader in AMI and smart grid for customer engagement. (ISGAN, 2014, p. 98)

ISGAN similarly describes the national Telegestore digital metering program in Italy as an innovative program that was a forerunner of things to come, allowing savings of 500 million Euro per year (ISGAN, 2019). Telegestore was "a revolution, not only in the technology, but also in the business processes, starting from the relationship with customers" (ibid.).

It is notable how often particular smart grid experiments are discussed; the same ones keep appearing in reports and reviews (see also Case Study 2.1). In this way, a handful of early smart grid experiments have shaped policy narratives about smart grids. Certain international organisations such as ISGAN and Mission Innovation, as well as the energy sector media, have played a central role in interpreting and framing these smart grid projects in positive ways. They have shaped the social memory of early smart grid projects, creating a form of false nostalgia (Routledge, 2017).

Smart grids nostalgia reflects with sentimentality on the early smart grid trials. Its storyline is that 'smart grids are do-able and bring lots of benefits—move early and you can be like these places, you can be internationally renowned'. This storyline provides positive motivation for other places to take action and to look on these early examples with admiration and a desire to replicate them. However, it also creates some problems because these pioneering smart grid experiments did have set backs. But the formal documented case studies of these experiments that populate reports and smart grid reviews have omitted these less positive aspects in favour of the nostalgic positive narrative (see also Case Study 4.2).

For instance, Pacific Gas and Electric (PG&E) utility in California had a high number of customer complaints in response to its smart metering program, which commenced in 2007. These complaints rose to such a high level that the Californian Public Utility Commission set up an inquiry to investigate, which identified a number of problems with the approach of PG&E, including:

> PG&E processes did not address the Customer concerns associated with the new equipment and usage changes. (Structure Consulting Group, 2010, p. 9)

> PG&E's system tolerances related to billing quality control were not stringent enough, resulting in multiple bill cancelations and re-billings, which were confusing to Customers. (ibid., p. 9)

The PG&E smart grid program cost considerably more than other Californian utilities, and the benefits were no way near as high. The costs of PG&E's program were reported as US$831 Million and benefits only US$19.6 Million (CPUC, 2016). By comparison, the other two utilities in California had higher benefits than costs for their smart grid programs.

The lack of widespread reporting of problems with pioneering smart grid programs leads to false expectations about the ease of implementing smart grid programs and projects—it is made to seem easier than it actually is—and learning from these pioneering energy innovations is hampered because the things that did not work are not talked about. However, as discussed in Chap. 4, usually, it is the things that did not work that we can learn the most from.

CASE STUDY 5.2 SCARCE DATA AND OFF-GRID HOUSEHOLDS IN AUSTRALIA

Australia is one of a group of countries at the forefront internationally of off-grid households because of its high penetration of rooftop solar and a very stringy grid (ENA, n.d.). Over a fifth of households in Australia now have solar installed: the highest proportion in the world (Australian PV Institute, 2019). The rate of rooftop solar uptake in Australia has been identified as a significant factor in the declining demand for electricity from the grid (AEMO, 2014, p. 5). Between 2009 and 2016, demand from the grid within Australia's National Electricity Market fell by 8% and is now expected to stay flat for the next twenty years (ibid., p. 4). The high rate of adoption of residential rooftop solar is anticipated to be repeated in the uptake of battery storage, with more efficient and affordable battery storage increasingly available in Australia. Over 31,000 batteries were installed by households in Australia in 2020, a 20% increase on installations in the previous year and up from 6750 in 2016 (Sunwiz, 2018; Vorrath, 2021).

However, even in Australia, with these fast-moving household changes, there is little data on who is off-grid. Data is not being actively collected on off-grid households because it is not thought to be happening at any significant scale. Energy utilities and other energy sector decision makers who nostalgically focus on the past and the way things used to be done, that is, supplying households with electricity from the grid, are gathering data on the electricity grid, but not on other types of electricity generation and supply. Australian studies of off-grid households are predominately about the economics of moving off-grid (Graham et al., 2015; Szatow & Moyse, 2014). The emphasis has been on modelling future scenarios but based on assumptions rather than actual data (see e.g., Brinsmead et al., 2015; Clean Energy Council, 2015; CSIRO, 2013). Off-grid households are an instance of *scarce data* (a contrast to the dominance of *big data* in modern society).

Along with a colleague (see Lovell & Watson, 2019; Fig. 5.2), I researched the availability of data on off-grid households in the State of Tasmania, Australia, to explore whether there was sufficient data to answer this question: how many households are currently off-grid in Tasmania? We found a wide variety of estimates of how many households were already off-grid in the State, ranging from 200 to 10,000. Our findings suggest considerable uncertainty in the data on off-grid households and an overall lack of data.

Our research also showed that a new generation of households is moving off-grid primarily for financial reasons (Lovell & Watson, 2019). For instance, one householder described how:

> we basically have free power. The initial cost was what we were going to spend anyway [on connection to the grid]. There are no ongoing costs for us, other than battery replacement at some point down the track, as all off-grid systems will have to do at some point. (Interview, October 2015)

Our research on Tasmania suggests that the framing of off-grid data collection in Australia remains nostalgically centred on the existing utilities and large-scale centralised energy infrastructure. An absence of data is a problem for energy sector innovation. It acts as a barrier to effective

Fig. 5.2 Image of solar panels at an off-grid home in Tasmania, Australia. (Source: Heather Lovell, University of Tasmania)

governance: the issue is not visible and is therefore not discussed by policy makers. Policy making is skewed towards governing data-rich policy areas. In the energy sector, this favours existing energy institutions, technologies and cultures, creating inertia and nostalgia, and making radical innovation difficult to achieve (Hughes, 1983). The role of data in creating opportunities for change is important, as the social scientists and critical data scholars Kitchin and Lauriault (2014, p. 4, emphasis added) explain:

> data are constitutive of the ideas, techniques, technologies, people, systems and contexts that conceive, produce, process, manage, and analyze them… **Data do not pre-exist their generation; they do not arise from nowhere and their generation is not inevitable**: protocols, organisational processes, measurement scales, categories, and standards are designed, negotiated and debated, and there is a certain messiness to data generation.

There have been several new energy sector policy initiatives in Australia since 2015 which recognise the general problem of lack of data visibility. These are not about off-grid data specifically but nevertheless provide some indication of increased policy attention to what we might call nostalgic data gaps. One example is the 2017 *Independent Review of the Future Security of the National Electricity Market in Australia* which placed a high priority on improving distributed energy resources (DER) data, including battery storage, explaining that:

> At present, AEMO [the Australian electricity market operator] lacks sufficient visibility of DER, which makes it difficult to manage the power system effectively. (Finkel, 2017, p. 32)

> The uptake of new technologies is putting residential, commercial and industrial consumers at the centre of the electricity market... [DER] such as rooftop solar photovoltaic and battery storage systems ... can all be harnessed to improve the reliability and security of the electricity system. **Improved access to data is needed to assist consumers, service providers, system operators and policy makers**. (ibid., p. 137)

The Australian electricity market operator (AEMO) has since set up a national DER register; from March 2020, all new DER connected to the electricity grid in Australia must be registered (AEMO, 2021). Some Australian energy utilities are starting to support off-grid households and collect off-grid data. For instance, Ergon Energy provides information on its website to support households making decisions about staying on or leaving the grid (Ergon Energy, 2018b). Horizon Power and Western Power are conducting trials supporting a small number of households to leave the grid in remote edge-of-grid areas in Western Australia (Ergon Energy, 2018a; Horizon Power, 2018).

Nostalgia can promote the continuation of past ways of doing things, often referred to as path dependency (Berkhout, 2002), which hampers innovation. In the case of off-grid households, a nostalgic approach to collecting data about the electricity grid might well be blinding us to significant off-grid changes already happening in Australia.

CASE STUDY 5.3 NOSTALGIA FOR BIG INFRASTRUCTURE: TENSIONS IN PLANNING FOR THE FUTURE OF THE GRID IN AUSTRALIA

In Australia, there are many different ideas about the future of the electricity grid. Some see the future of the grid as being splintered, decentralised, populated by smaller micro grids and many more off-grid industries and households (see e.g., AECOM, 2014; Szatow & Moyse, 2014). Another version of the future, and a version that has dominated mainstream government and utility planning in Australia over the past few years, sees the electricity grid as not only remaining in place but strengthened. In this case study I focus on the latter version of the future—a future with greater interconnection—by examining the role of nostalgia in energy system planning in the Australian State of Tasmania.

The island State of Tasmania has been connected to mainland Australia with an undersea electricity cable (the Basslink) since 2006. In recent years, there has been planning and discussion about adding a second undersea cable (the Marinus Link) to boost Tasmania's capacity to export its renewable electricity (TasNetworks, 2021). Tasmania generates 100% of its electricity by renewable energy (mostly generated by large-scale hydroelectricity plants, the majority of which were built in the 1940s and 1950s). There are also plans to build new pumped hydroelectricity ('pumped hydro') storage plants in Tasmania, using the existing infrastructure to provide large-scale electricity storage. These pumped hydro plans are tied up with Marinus Link because both rest on the idea of Tasmania being able to provide more energy services to mainland Australia through enhanced electricity grid transmission. As another layer to all this big infrastructure planning in Tasmania, the state government recently announced Tasmania's intention to double its renewable electricity generation by 2040: an extra 10,500 GWh per year (Tasmanian Department of State Growth, 2020).

Tasmania's plans rely on the continuation of a centralised electricity grid across Australia, that is, business as usual within Australia's electricity market, the NEM. Indeed, Tasmania is closely following the NEM Integrated System Plan developed by AEMO (see Case Study 3.2). Within AEMO's Integrated System Plan, there is an emphasis on building stronger electricity interconnections between the Australian states that are part of the NEM, as it explains:

The projected portfolio of new resources involves substantial amounts of geographically dispersed renewable generation, placing a greater reliance on the role of the transmission network. **A much larger network footprint with transmission investment will be needed to efficiently connect and share these low fuel cost resources.** (AEMO, 2018, p. 6, emphasis added)

The State of Tasmania is one of the NEM states that is planning for greater interconnection. The estimated cost of Project Marinus is A\$3.5 billion (TasNetworks, 2020, p. 5), so if the centralised grid system in Australia erodes, then this big infrastructure investment will have been a very costly mistake. Possibly it is worth taking this risk, as there are significant opportunities for Tasmania if electricity storage is required at a large scale by the grid several years hence, when as much as 12,000 MW of coal-fired power station capacity will be retired (TasNetworks, 2020, p. 2). A key question though is whether Tasmania's planning is being driven, at its heart, by nostalgia. Clearly, there are several practical considerations around how best to manage uncertainty and disruption in planning for Australia's energy future, and, given the large sunk costs in existing energy infrastructure, it makes sense in lots of ways to continue investing more in this existing infrastructure. But is there also a more emotional, nostalgic aspect of this 'big infrastructure' planning? There are a number of signs that this might be the case. First is the absence of any other versions of the future in the Tasmanian Renewable Energy Plan (e.g., a decentralised grid, a self-sufficient energy island). Throughout the Tasmanian Renewable Energy Plan, it is assumed that a strong, national transmission system will be important to Australia in the future, and no other versions of the future are considered. Second, there is a clear nostalgic evoking of the past in the Plan, with the benefits of historical large-scale investment in hydroelectric infrastructure mentioned frequently, for instance:

Tasmania has **a long history of major industrial development powered by renewable energy** and there are major opportunities for the establishment of jobs-rich, large-scale, energy intensive enterprises in the state (Tasmanian Department of State Growth, 2020, p. 45, emphasis added)

This indicates nostalgic thinking in relation to the electricity grid. The historical renewable energy referred to in the quotation above is Tasmania's extensive network of thirty hydroelectricity power stations, which were developed mostly in the period 1930 to 1960 (Hydro Tasmania, 2014).

The resulting plentiful and relatively low cost electricity attracted a number of large international industries to the State, including a zinc smelter, pulp and paper mill and aluminium smelter. These large industries still reside in Tasmania, and constitute around half of Tasmania's electricity demand. The nostalgia stems primarily from this era of hydroelectricity development and subsequent economic growth. It is a form of *institutional* nostalgia, with the government-owned utilities and the state government promoting a version of the future in which the electricity sector once again acts as the principal driver of economic development.

In Tasmania's planning for Project Marinus and other energy infrastructure investment there is an underlying expectation that the electricity grid will continue as in the past, but with even more investment in infrastructure—a *gold plating* of the grid so that it keeps up with a host of disruptive innovations such as decentralised generation and storage. This version of the future might indeed come to pass. However, it is a version of the future that is heavily path dependent, and history tells us that infrastructure systems do periodically have radical breaks from the past (Bridge et al., 2018; Hughes, 1983). For example, one alternative scenario is that the Australian states revert back to more state-centred electricity systems, rather than focusing so much on investment in the NEM. Australian states have sovereignty over their energy systems and the NEM operates on a consensus model, therefore its authority could be eroded. AEMO's Integrated System Plan promotes the efficiency of sharing electricity across state boundaries, but there is an alternative future wherein the Australian states revert to operating more in isolation, because increasingly they have stronger technological capability to do so, as renewable generation and battery storage increase (an example of this is the South Australia's Hornsdale Power Reserve (see ARENA, 2020)).

Learning from Smart Grids and Nostalgia

Nostalgia can influence contemporary energy sector innovation through its promotion of, and attachment to, sentimental or romanticised versions of the past. As we see in the case studies presented in this chapter, it can prevent us from learning from the mistakes and failures of past energy innovation (pioneering smart grid experiments) and encourage potentially risky future energy innovation by prioritising past desires and needs (large-scale transmission infrastructure). In the table below, I summarise the key learnings from these case studies and suggest how they might guide future practice.

Key learning	Recommendation for energy practitioners
The past has influence in all sorts of ways in energy, not just in terms of the technical legacy of infrastructure, but also cultural and emotional attachments to ways of seeing things and what is judged to be important. This influence affects what problems are identified and the types of solutions proposed. Nostalgia helps explain why many innovations are not initially seen or made visible and why there might be a reluctance to engage with them.	Consider how the centralised grid version of the future is heavily influenced by nostalgia. In other words, it is about maintaining an existing way of providing electricity, promoting a continuation of infrastructure, flows of capital and organisations. It is a risky strategy because it involves high investment up front in the face of a very uncertain future, with declining electricity demand from the grid and an increase in a host of new decentralised technologies. Caution should be applied to backing this version of the future, which is judged to be driven at least in part by nostalgia. This situation can result in scarce data on fast changing aspects of the energy sector. Scarce data acts as a barrier to effective governance and results in energy policy making skewed towards governing data-rich policy areas. High-level mapping of existing and future energy activities, technologies and processes and their associated data should be undertaken to identify instances of scarce data. This could be conducted by existing energy data organisations (such as CSIRO Data 61 and AEMO in the Australian context).

REFERENCES

AECOM. (2014). *Australia's off-grid clean energy market.* Research Paper for ARENA.

AEMO. (2014, June). *National Electricity Forecasting Report 2014.* Australian Energy Market Operator (AEMO). Retrieved August 12, 2019, from http://www.aemo.com.au/Electricity/Planning/Forecasting/National-Electricity-Forecasting-Report

AEMO. (2018). *2018 Integrated System Plan for the National Electricity Market.* Australian Energy Market Operator (AEMO). Retrieved June 16, 2020, from https://aemo.com.au/-/media/files/electricity/nem/planning_and_fore-casting/isp/2018/integrated-system-plan-2018_final.pdf

AEMO. (2021). *Distributed energy resource register.* Australian Energy Market Operator (AEMO). Retrieved March 31, 2021, from https://aemo.com.au/en/energy-systems/electricity/der-register

ARENA. (2020). *South Australian battery grows bigger and better.* Retrieved June 15, 2021, from https://arena.gov.au/blog/south-australian-battery-grows-bigger-and-better/

Atia, N., & Davies, J. (2010). Nostalgia and the shapes of history. *Memory Studies, 3*(3), 181–186.

Australian PV Institute. (2019). *PV in Australia Report 2018*. Australian PV Institute (APVI). Retrieved March 30, 2021, from http://apvi.org.au/wp-content/uploads/2019/10/NSR-Guidelines-2018_AUSTRALIA_v2.pdf

Berkhout, F. (2002). Technological regimes, path dependency and the environment. *Global Environmental Change, 12*(1), 1–4.

Bridge, G., Barr, S., Bouzarovski, S., Bradshaw, M., Brown, E., Bulkeley, H., & Walker, G. (2018). *Energy and society: A critical perspective*. Routledge.

Brinsmead, T. S., Graham, P., Hayward, J., Ratnam, E. L., & Reedman, L. (2015). *Future energy storage trends: An assessment of the economic viability, potential uptake and impacts of electrical energy storage on the NEM 2015–2035*. CSIRO Report Number EP155039. CSIRO.

Cashman, R. (2006). Critical nostalgia and material culture in Northern Ireland. *Journal of American Folklore, 119*, 137–160.

Clean Energy Council. (2015). *Australian energy storage roadmap*. Clean Energy Council.

CPUC. (2016). *California Smart Grid Annual Report to the Governor and the Legislature*. California Public Utilities Commission (CPUC). Retrieved March 13, 2021, from https://www.cpuc.ca.gov

CSIRO. (2013). *Change and choice: The Future Grid Forum's analysis of Australia's potential electricity pathways to 2050*. CSIRO Future Grid Forum. Retrieved June 13, 2016, from https://publications.csiro.au/rpr/download?pid=csiro:EP1312486&dsid=DS13

Czarniawska, B. (1997). *Narrating the organization: Dramas of institutional identity*. University of Chicago Press.

Dames, N. (2010). Nostalgia and its disciplines: A response. *Memory Studies, 3*(3), 269–275.

Davies, J. (2010). Sustainable nostalgia. *Memory Studies, 3*(3), 262–268.

ENA. (n.d.). *Guide to Australia's Energy Networks*. Energy Networks Australia (ENA).

Ergon Energy. (2018a). *Get $50 cashback*. Retrieved October 12, 2018, from https://www.ergon.com.au/network/smarter-energy/battery-storage

Ergon Energy. (2018b). *Going off-grid*. Retrieved October 25, 2018, from https://www.ergon.com.au/network/smarter-energy/battery-storage/going-off-grid

Finkel, A. (2017). *Independent Review into the Future Security of the National Electricity Market: Blueprint for the Future*. Commonwealth of Australia.

Graham, P., Brinsmead, T., Reedman, L., Hayward, J., & Ferraro, S. (2015). *Future Grid Forum – 2015 Refresh: Technical report*. CSIRO report for the Energy Networks Association, Australia.

Hanel, T., & Hård, M. (2015). Inventing traditions: Interests, parables and nostalgia in the history of nuclear energy. *History and Technology, 31*(2), 84–107.

Horizon Power. (2018). *With State Government support, Horizon Power is transforming the traditional electricity network into a highly distributed intelligent microgrid*. Retrieved October 25, 2018, from https://horizonpower.com.au/our-community/projects/

Hughes, T. P. (1983). *Networks of power: Electrification in Western society 1880–1930*. The John Hopkins University Press.

Hydro Tasmania. (2014). *The power of nature*. Hydro Tasmania. https://www.hydro.com.au/docs/default-source/clean-energy/our-power-stations/power-of-nature.pdf?sfvrsn=15c22528_2

ISGAN. (2014). *AMI Case Book Version 2.0: Spotlight on Advanced Metering Infrastructure*. International Smart Grid Action Network (ISGAN).

ISGAN. (2019). *AMI CASE Case05 – ITALY*. International Smart Grid Action Network (ISGAN). Retrieved September 10, 2019, from http://www.iea-isgan.org/ami-case-case05-italy/

Kitchin, R., & Lauriault, T. P. (2014). Towards critical data studies: Charting and unpacking data assemblages and their work. In J. Eckert, A. Shears, & J. Thatcher (Eds.), *Geoweb and Big Data*. University of Nebraska Press.

Lovell, H., & Watson, P. (2019). Scarce data: Off-grid households in Australia. *Energy Policy, 129*, 502–510.

Pickering, M., & Keightley, E. (2006). The modalities of nostalgia. *Current Sociology, 54*(6), 919–941.

Routledge, C. (2015). *Nostalgia: A psychological resource*. Routledge.

Routledge, C. (2017). *Approach with caution: Nostalgia is a potent political agent*. Retrieved February 14, 2021, from https://undark.org/2017/10/31/nostalgia-power-politics-trump/

Sedikides, C., Wildschut, T., Arndt, J., & Routledge, C. (2008). Nostalgia: Past, present, and future. *Current Directions in Psychological Science, 17*(5), 304–307.

Structure Consulting Group. (2010). *PG&E Advanced Metering Assessment Report*. Commissioned by the California Public Utilities Commission (CPUC). Retrieved June 12, 2019, from https://www.pge.com/includes/docs/pdfs/myhome/customerservice/meter/smartmeter/StructureExecutiveSummary.pdf

Sunwiz. (2018). *Australian Battery Market trebles in 2018*. Retrieved April 1, 2018, from http://sunwiz.com.au/index.php/2012-06-26-00-47-40/73-newsletter/434-australian-battery-market-trebles-in-2018.html

Szatow, T., & Moyse, D. (2014). *What happens when we unplug? Exploring the consumer and market implications of viable, off-grid energy supply*. Energy for the People and the Australian Technology Association (ATA).

Tasmanian Department of State Growth. (2020). *Tasmanian Renewable Energy Action Plan*. Tasmanian Government. Retrieved December 20, 2020, from https://renewablestasmania.tas.gov.au/__data/assets/pdf_file/0008/275876/Tasmanian_Renewable_Energy_Action_Plan_December_2020.pdf

TasNetworks. (2020). *Marinus Link – Delivering low cost, reliable and clean energy*. Retrieved March 12, 2021, from https://www.marinuslink.com.au/wp-content/uploads/2020/01/Project-Marinus-Fact-Sheet.pdf

TasNetworks. (2021). *Marinus Link*. Retrieved March 31, 2021, from https://www.marinuslink.com.au/

Van House, N., & Churchill, E. F. (2008). Technologies of memory: Key issues and critical perspectives. *Memory Studies, 1*(3), 295–310.

Vorrath, S. (2021). *Australians installed 31,000 batteries in 2020, led by households*. Retrieved March 31, 2021, from https://onestepoffthegrid.com.au/australians-installed-31000-batteries-in-2020-led-by-households/

Whitehead, H. (2010). The agency of yearning on the Northwest Coast of Canada: Franz Boas, George Hunt and the salvage of autochthonous culture. *Memory Studies, 3*(3), 215–223.

CHAPTER 6

Conclusions

Understanding energy innovation has shown energy innovation to be a messy process—a complex mix of technological advances, politics, and social learning and adaptation. A wide range of people and things are involved in energy innovation, from electricity meters to households. A downside of many academic theories is that they isolate one particular aspect of the innovation process and study that aspect to the exclusion of other processes that might be equally important. What I have tried to do in this book is to give an overview of many different types of theory, to show how these concepts and ideas might be applied together and in different contexts, to help further our understanding of energy innovation. I do so by using a range of smart grid case studies and grouping research findings under the four themes of nodes, networks, narratives, and nostalgia.

Not everything, of course, neatly fits into these four themes, so I am also guilty of trying to *tidy up* energy innovation. There is an underlying tension between abstracting and conceptualising energy innovation. Conceptualisation helps our understanding by identifying core actors and processes, but at the same time, it risks oversimplifying, either by missing the rich detail of each individual case study or by seeing everything through one particular conceptual lens. In this short concluding chapter, each of the four chapters is briefly summarised, including the key ideas and learnings, followed by a reflection on smart grids and energy innovation.

© The Author(s) 2022
H. Lovell, *Understanding Energy Innovation*,
https://doi.org/10.1007/978-981-16-6253-9_6

KEY IDEAS AND LEARNINGS

Networks

There are many different types of network that social scientists study to better understand processes of change, from policy networks to sociotechnical networks. In Chap. 2, I explored what is usefully highlighted when we conceptualise the different technologies and people involved in energy innovation as networks, using the case study of smart grids to consider all the types of actor involved. I looked at three short case studies of networks: international smart grid policy networks; a local community network on Bruny Island, Australia; and a fragile network—a digital metering program in the State of Victoria, Australia.

Networks are present at all sorts of scales and with different types of substance binding them together. The key features of networks that make it a useful metaphor for energy innovation are interconnectedness (relationships), flows, network-wide effects, and fragility. A lot of work goes into keeping networks stable; they are inherently fragile things, prone to breaking down. The electricity grid is a good example of this. A huge amount of resources (time, people, expertise, and technology) is applied constantly behind the scenes to keep our electricity supply running. It is only on rare occasions when the electricity grid breaks down that these resources are exposed and made visible, and the inherent fragility of the electricity network is revealed.

Some key learnings about energy innovation from the study of networks are:

- There are lots of different types of network relevant to energy innovation: policy, social, sociotechnical, and business.
- Energy programs and new initiatives, such as smart grids, are sometimes misconceived as technical programs, whereas in reality, they are sociotechnical (i.e., part social and part technical).
- Decisions about energy innovations in any particular locale (state, city, region) are not made in isolation. International policy networks continuously circulate new ideas and information, and these information flows can have both positive and negative effects.
- Well-functioning energy networks may appear stable, but actually, they are quite fragile: there is a lot of work going on behind the scenes to give the illusion of stability.

Nodes

Nodes play an important role in providing stability—keeping things the same—as well as innovating. In Chap. 3, I analysed the role of nodes within energy innovation. Nodes are stable points on networks, points of intersection. In three case studies, I focused on very different types of smart grid social and technical node: the digital electricity meter, an energy sector organisation (the Australian Energy Market Operator), and an island.

Thinking about energy sector components and organisations as nodes is a more static conceptualisation compared with a network conceptualisation. Whereas the network metaphor encourages us to think about flows, the idea of nodes instead focuses our attention on the organisations, people, and technologies that provide anchor points, and often act as key brokers at significant junctions within processes of energy innovation. In the case of the digital electricity meter, nodes are technologies at the intersection of households and the grid. Nodes are also organisations that play important roles in running the energy sector (the national regulators and rule setters) and in international energy research and knowledge dissemination (e.g., the International Smart Grid Action Network), as well as standard setting (e.g., the International Electrotechnical Commission). Households can also usefully be thought of as nodes; central actors who not only consume energy but also increasingly generate it from rooftop solar and store it in household batteries. Nodes can also be influential individuals, such as entrepreneurial brokers who innovate in technology and/or policy.

The concept of nodes is useful because it enables us to concentrate our analysis on the critical components of energy innovation and better understand the work these components do—whether they are social or technical—in keeping things the same and innovating. A focus on nodes also helps us understand what happens when things go wrong, as it is often that a key node has broken down or is no longer working in the way it used to.

Some key learnings about energy innovation from the study of nodes are:

- Nodes have a strong influence over energy innovation because of their role in managing and co-ordinating flows in networks, and hence are worthy of attention.

- Nodes typically have what is termed interpretative flexibility, that is, they are understood differently by different actors. This flexibility is generally seen as a strength allowing them to function.
- Attempts are often made to replicate successful nodes elsewhere, in different contexts, but this does not always work because the things and people they are co-ordinating are different.
- Because nodes are embedded within their networks, they are the product of flows within those networks. Nodes are therefore less adept at recognising and driving change outside of the network in which they are situated, that is, in effecting more radical innovation. When nodes are positioned at the intersection of different networks (e.g., policy networks), they are particularly active and influential.

Narratives

The study of narratives is important in helping us to simplify and make sense of innovations, including in the energy sector. In Chap. 4, I analysed three examples of narratives about smart grids: a global industry narrative about households and their willingness to participate in smart grids; a narrative of policy failure about a particular smart grid project in the State of Victoria, Australia; and the narratives that currently compete with smart grids in Australia, including the hydrogen economy and off-grid energy futures.

Social research tells us that we understand and make sense of the world through stories. From studies of scientists in their labs to ethnographies of government and corporations, narratives have been found to underpin, structure, and give meaning to the work that we do (Czarniawska, 2004). It is no different for the energy sector, where stories circulate about successes and failures, about particular technologies and policy experiments. Narratives are a way of simplifying the messiness of innovation and change processes and making them more readily understandable, distilling the main points. But through this simplification, narratives can also be dangerous: the need for a coherent story means that significant details are often left out. So, for example, a story about a smart grid project failure has no room in it for the successes of that project, as it detracts from the narrative.

Narratives are particularly important in situations of newness, ambiguity and uncertainty (Hajer, 1995). These are characteristic of much contemporary energy sector innovation, which involves lots of new technologies and new modes of operation. This innovation is occurring within a sector where there has not been significant change for several decades in terms of the principal mode of electricity generation, distribution, and consumption. So, paying attention to narratives, as well as to their silences and gaps—what has been left out—is one way to develop a better understanding of societal responses to energy innovation.

Some key learnings about energy innovation from the study of narratives are:

- All narratives have particular framings of the policy problem and, therefore, its solutions.
- Within every narrative, there are silences—things that are left out— sometimes deliberately and sometimes accidentally and these are worth noticing.
- Some narratives become very popular because they are a good strategic fit, and organisations with vested interests drive the narrative, but there may actually be little evidence to substantiate the narrative (e.g., the willing prosumer narrative in Australia).
- Learning from energy sector failure is more difficult than learning from success because there is much less information circulating about failures.

Nostalgia

Nostalgia is a focus on and longing for the past and past ways of doing things—a sentimental feeling that it would be nice if things were as they were previously. In Chap. 5, I looked at how nostalgia can hamper efforts at energy innovation, both in terms of how it blinds us to change already under way and how memories of things and ways of doing can encourage or hinder innovation. I drew on three diverse case studies: memories of pioneering international smart grid experiments and their present-day effect; data about off-grid households in Australia; and nostalgia for big infrastructure energy solutions in Australia.

Applying ideas about nostalgia to energy innovation might seem rather odd at first glance. However, it is useful in showing us how certain ways of doing and particular expectations are still focused on the past rather than thinking about the future of the energy sector. Focus on the past creates a situation where new areas of innovation and change are not paid as much regard as they could be. New ways of doing are simply not on the radar, as contemporary systems and organisations continue to do things the way they have always done them. For instance, data is not being collected on households leaving the electricity grid in Australia because this has not happened much in the past, so contemporary data collection systems do not recognise off-grid households.

Some key learnings about energy innovation from the study of nostalgia are:

- The past has influence in all sorts of ways in energy, not just in terms of the technical legacy of infrastructure, but also cultural and emotional attachments to ways of seeing things and what is judged to be important.
- The influence of nostalgia affects what problems are identified and the types of solutions proposed.
- Nostalgia helps explain why many innovations are not initially seen or made visible and why there might be a reluctance to engage with them, for example leading to situations of scarce data.

REFLECTION: UNDERSTANDING ENERGY INNOVATION THROUGH INTEGRATING THE SOCIAL

There has been lots of attention to energy sector innovation in recent years. The energy sector is grappling with a range of problems, from climate change to increased consumer-led investment in distributed energy resources, to the opportunities afforded by new digital technologies and data. Terms such as energy transition and roadmaps are in frequent use world-wide (Clean Energy Council, 2015; South Korean Government, 2019). There are many possible pathways ahead. We are likely to see increased diversity in forms of energy provision around the world, as different choices are made about energy futures beyond centralised fossil fuel provision. These changes are complex. However, one certain feature is that energy sector innovation will continue to comprise a mix of the social and the technical. Seemingly technical decisions about transmission lines and renewable energy zones have social, political, and cultural dimensions at their core. Social domains such as household preferences and habits are heavily influenced by technology, for example, how much the technology meets their needs, how easy it is to understand, and how reliable it is in its function.

To take one example from the book, the case study of international policy networks of smart grid innovation comprising entrepreneurs, companies, governments, smart grid projects, and technologies (see Case Study 2.1, Chap. 2). Smart grids are a global phenomenon, and I demonstrate how these international networks affect what happens on the ground in any place something *smart* is happening with the grid. The effect of smart grid projects being monitored globally through these networks is worth paying attention to because it is influential in determining energy innovation processes. Whatever the type of energy innovation, wherever it is, there is a huge amount of effort that goes into making it appear to be successful, as this grants a place on the international stage to the people involved and their locality. But, as we well know, not everything works in attempts at energy innovation. Every project needs smoothing over and attention; this type of smoothing over often comes at the expense of wider learning from mistakes.

International policy networks are just one example of how developing a better understanding of energy innovation necessitates paying attention to both the social and the technical elements of innovation and their many twists and turns. Smart grids provide a good illustration of how interwoven the social and the technical are. At first glance, smart grids are a straightforward technical solution to a number of pressing energy sector policy problems. In reality, smart grids are as much social as they are technical. It is my hope that *Understanding Energy Innovation* convincingly shows this to be the case.

References

Clean Energy Council. (2015). *Australian energy storage roadmap*. Clean Energy Council.

Czarniawska, B. (2004). *Narratives in social science research*. Sage.

Hajer, M. A. (1995). *The politics of environmental discourse: Ecological modernisation and the policy process*. Clarendon Press.

South Korean Government. (2019). *Roadmap to the worlds' best leading hydrogen economy*. South Korean Department of Energy New Industry. Retrieved July 13, 2020, from http://www.motie.go.kr/motie/ne/presse/press2/bbs/bbsView.do?bbs_cd_n=81&cate_n=1&bbs_seq_n=161262

Index